CAMBRIDGE LIBRARY COLLECTION

Books of enduring scholarly value

Botany and Horticulture

Until the nineteenth century, the investigation of natural phenomena, plants and animals was considered either the preserve of elite scholars or a pastime for the leisured upper classes. As increasing academic rigour and systematisation was brought to the study of 'natural history', its subdisciplines were adopted into university curricula, and learned societies (such as the Royal Horticultural Society, founded in 1804) were established to support research in these areas. A related development was strong enthusiasm for exotic garden plants, which resulted in plant collecting expeditions to every corner of the globe, sometimes with tragic consequences. This series includes accounts of some of those expeditions, detailed reference works on the flora of different regions, and practical advice for amateur and professional gardeners.

An Historical Account of Coffee

This tract, which first appeared in 1774, considers the characteristics, cultivation and uses of the coffee plant. Its author, John Ellis (*c.*1710–76), was a botanist and zoologist who from 1770 to 1776 served as a London agent for the government of Dominica. Published in order to promote the prosperity of the island, the work reflects the difficulties faced by the coffee growers. Ellis begins by describing the flower and fruit of the coffee plant. He then presents his historical survey, drawing on contemporaneous travel writing to illuminate coffee-related practices around the globe. The narrative takes in the plant's early uses in Arabia, its cultivation in the colonies, and the growth of coffee houses in Europe. This reissue also contains a 1770 work by Ellis which gives instructions on transporting plants overseas. Reissued elsewhere in this series is *The Early History of Coffee Houses in England* (1893).

T0225011

Cambridge University Press has long been a pioneer in the reissuing of out-of-print titles from its own backlist, producing digital reprints of books that are still sought after by scholars and students but could not be reprinted economically using traditional technology. The Cambridge Library Collection extends this activity to a wider range of books which are still of importance to researchers and professionals, either for the source material they contain, or as landmarks in the history of their academic discipline.

Drawing from the world-renowned collections in the Cambridge University Library and other partner libraries, and guided by the advice of experts in each subject area, Cambridge University Press is using state-of-the-art scanning machines in its own Printing House to capture the content of each book selected for inclusion. The files are processed to give a consistently clear, crisp image, and the books finished to the high quality standard for which the Press is recognised around the world. The latest print-on-demand technology ensures that the books will remain available indefinitely, and that orders for single or multiple copies can quickly be supplied.

The Cambridge Library Collection brings back to life books of enduring scholarly value (including out-of-copyright works originally issued by other publishers) across a wide range of disciplines in the humanities and social sciences and in science and technology.

An Historical Account of Coffee

With an Engraving,
and Botanical Description of the Tree

JOHN ELLIS

CAMBRIDGE
UNIVERSITY PRESS

CAMBRIDGE
UNIVERSITY PRESS

University Printing House, Cambridge, CB2 8BS, United Kingdom

Published in the United States of America by Cambridge University Press, New York

Cambridge University Press is part of the University of Cambridge.
It furthers the University's mission by disseminating knowledge in the pursuit of
education, learning and research at the highest international levels of excellence.

www.cambridge.org
Information on this title: www.cambridge.org/9781108066884

© in this compilation Cambridge University Press 2013

This edition first published 1774
This digitally printed version 2013

ISBN 978-1-108-06688-4 Paperback

Selected botanical reference works available in the
CAMBRIDGE LIBRARY COLLECTION

al-Shirazi, Noureddeen Mohammed Abdullah (compiler), translated by Francis Gladwin: *Ulfáz Udwiyeh, or the Materia Medica* (1793) [ISBN 9781108056090]

Arber, Agnes: *Herbals: Their Origin and Evolution* (1938) [ISBN 9781108016711]

Arber, Agnes: *Monocotyledons* (1925) [ISBN 9781108013208]

Arber, Agnes: *The Gramineae* (1934) [ISBN 9781108017312]

Arber, Agnes: *Water Plants* (1920) [ISBN 9781108017329]

Bower, F.O.: *The Ferns (Filicales)* (3 vols., 1923–8) [ISBN 9781108013192]

Candolle, Augustin Pyramus de, and Sprengel, Kurt: *Elements of the Philosophy of Plants* (1821) [ISBN 9781108037464]

Cheeseman, Thomas Frederick: *Manual of the New Zealand Flora* (2 vols., 1906) [ISBN 9781108037525]

Cockayne, Leonard: *The Vegetation of New Zealand* (1928) [ISBN 9781108032384]

Cunningham, Robert O.: *Notes on the Natural History of the Strait of Magellan and West Coast of Patagonia* (1871) [ISBN 9781108041850]

Gwynne-Vaughan, Helen: *Fungi* (1922) [ISBN 9781108013215]

Henslow, John Stevens: *A Catalogue of British Plants Arranged According to the Natural System* (1829) [ISBN 9781108061728]

Henslow, John Stevens: *A Dictionary of Botanical Terms* (1856) [ISBN 9781108001311]

Henslow, John Stevens: *Flora of Suffolk* (1860) [ISBN 9781108055673]

Henslow, John Stevens: *The Principles of Descriptive and Physiological Botany* (1835) [ISBN 9781108001861]

Hogg, Robert: *The British Pomology* (1851) [ISBN 9781108039444]

Hooker, Joseph Dalton, and Thomson, Thomas: *Flora Indica* (1855) [ISBN 9781108037495]

Hooker, Joseph Dalton: *Handbook of the New Zealand Flora* (2 vols., 1864–7) [ISBN 9781108030410]

Hooker, William Jackson: *Icones Plantarum* (10 vols., 1837–54) [ISBN 9781108039314]

Hooker, William Jackson: *Kew Gardens* (1858) [ISBN 9781108065450]

Jussieu, Adrien de, edited by J.H. Wilson: *The Elements of Botany* (1849) [ISBN 9781108037310]

Lindley, John: *Flora Medica* (1838) [ISBN 9781108038454]

Müller, Ferdinand von, edited by William Woolls: *Plants of New South Wales* (1885) [ISBN 9781108021050]

Oliver, Daniel: *First Book of Indian Botany* (1869) [ISBN 9781108055628]

Pearson, H.H.W., edited by A.C. Seward: *Gnetales* (1929) [ISBN 9781108013987]

Perring, Franklyn Hugh et al.: *A Flora of Cambridgeshire* (1964) [ISBN 9781108002400]

Sachs, Julius, edited and translated by Alfred Bennett, assisted by W.T. Thiselton Dyer: *A Text-Book of Botany* (1875) [ISBN 9781108038324]

Seward, A.C.: *Fossil Plants* (4 vols., 1898–1919) [ISBN 9781108015998]

Tansley, A.G.: *Types of British Vegetation* (1911) [ISBN 9781108045063]

Traill, Catherine Parr Strickland, illustrated by Agnes FitzGibbon Chamberlin: *Studies of Plant Life in Canada* (1885) [ISBN 9781108033756]

Tristram, Henry Baker: *The Fauna and Flora of Palestine* (1884) [ISBN 9781108042048]

Vogel, Theodore, edited by William Jackson Hooker: *Niger Flora* (1849) [ISBN 9781108030380]

West, G.S.: *Algae* (1916) [ISBN 9781108013222]

Woods, Joseph: *The Tourist's Flora* (1850) [ISBN 9781108062466]

For a complete list of titles in the Cambridge Library Collection please visit:
http://www.cambridge.org/features/CambridgeLibraryCollection/books.htm

A   B   C   D   F   G   H   E

COFFEA *Arabica.*

J. Miller Sc.

The material originally positioned here is too large for reproduction in this
reissue. A PDF can be downloaded from the web address given on page iv
of this book, by clicking on 'Resources Available'.

AN

HISTORICAL ACCOUNT

OF

COFFEE.

WITH

An Engraving, and Botanical Defcription of the TREE.

TO WHICH ARE ADDED

Sundry Papers relative to its Culture and Ufe, as an Article of
DIET and of COMMERCE.

PUBLISHED by JOHN ELLIS, F. R. S.
AGENT FOR THE ISLAND OF DOMINICA.

LONDON:
PRINTED FOR EDWARD AND CHARLES DILLY.
MDCCLXXIV.

PREFACE.

THE objects of this performance are, the promotion of science, national advantage, and the prosperity of the Island for which I have the honour to be Agent. The description of Coffee, with the exact delineation of all its parts, together with the History of its introduction and progress, will contribute to the first. In respect to the two last, I own myself obliged to my friend Dr. Fothergill. The importance of giving encouragement to the growth of this article for home consumption, and exportation, had often been the subject of our conversation, and I begged he would seize some opportunity to give me his sentiments in writing. He allows me to insert them in this publication. Some other Papers and Letters relative to my design having been communicated to me by Gentlemen well versed in the subject, I think it proper, on this occasion, to lay them likewise before the Public;

and

and hope the Weſt India planter will find here ſome uſeful information, the Legiſlature convincing motives for putting his produce upon at leaſt as favourable a footing, with reſpect to duties, as foreign articles uſed for the ſame purpoſe. I ſhall eſteem myſelf very happy, if theſe endeavours to promote the advantage of my conſtituents, and of the community in general, ſhould meet with the deſired ſucceſs.

J. ELLIS.

A BOTA-

A BOTANICAL

DESCRIPTION

OF THE

FLOWER and FRUIT of the COFFEE-TREE.

THE characters of that genus of plants called COFFEA by Linnæus, who places it in the first order of his fifth class, that is, among the *Pentandria Monogynia*, or plants that have five male organs and one female organ, are as follows:

CALIX. *Perianthium* quadridentatum, minimum, germini insidens.

The *Flower Cup*, whose brim has four very small indentations, and is placed upon the germen or embryo seed vessel.

COROLLA. *Petalum* infundibuliforme. *Tubus* cylindraceus, tenuis, calyce multoties longior. *Limbus* planus, quinquepartitus, tubo longior; *laciniis* lanceolatis, lateribus revolutis.

The *Flower* consists of one funnel-shaped petal, with a slender tube nearly cylindrical, much longer than the flower cup. Its brim is expanded and divided into five

B   segments,

segments. Thefe are longer than the tube, are fharp-pointed, and reflexed on the fides.

STAMINA. *Filamenta* quinque, fubulata, tubo corollæ impofita. *Antheræ* lineares, incumbentes, longitudine filamentorum.

The *Chives* confift of five awl-fhaped filaments fixed on the tube of the flower. On thefe are placed the linear-fhaped fummits, containing the male duft. Thefe are of the fame length with the filaments.

PISTILLUM. *Germen* fubrotundum. *Stylus* fimplex, longitudine corollæ. *Stigmata* duo, reflexa, fubulata crafliufcula.

The *Piftil* confifts of a round-ifh germen, or embryo feed veffel. The ftyle is ftreight and even, of the length of the flower, and ends in two flender, reflexed, fpungy tops.

PERICARPIUM. *Bacca* fubrotunda, puncto umbilicata.

The *Fruit* is a roundifh berry, marked at the top with a puncture like a navel.

SEMINA. Bina, elliptico-hemifphærica, hinc gibba, inde plana, arillo involuta.

It has two feeds, of an oblong hemifpherical form, convex on the one fide, and flat on the other ; each of which is enclofed in a membrane, commonly called the parchment.

*** Linnæus has omitted taking notice of the feptum or membrane that divides the feeds into two cells or loculaments; and alfo the little furrow on the flat fide of each feed —It frequently happens that in the Mocha Coffee only one feed is to be found, the other being defective.

A fhort

A short Description of the COFFEE TREE, *taken from* Linnæus's Amœnitat. Academ. vol. VI. p. 169.

Arbor simplex, erecta, minus alta; *Ramis* longis, simplicibus, laxis & fere nutantibus, vestitis *Foliis* oppositis, laurinis, sempervirentibus, ornata *Floribus* albis sessilibus, fere Jasmini corolla, quibus *Baccæ* cerasorum facie rubicundæ succedunt, pulpâ pallidâ, submucilaginosâ, fatuâ, intus gerentes semina duo, dura, hinc convexa, inde plana, arillo cartilagineo vestita.

The *Tree* grows erect, with a single stem, is but low [*from eight to twelve feet high*], has long undivided, slender branches, bending downwards. These are furnished with evergreen opposite leaves, not unlike those of the bay tree, and adorned with white Jasmine flowers sitting on short foot-stalks, which are succeeded by red berries like those of the cherry, having a pale, insipid, glutinous pulp, containing two hard seeds, convex on the one side, and flat on the other, which are covered with a cartilaginous membrane or parchment.

This tree is a native of Arabia Felix, and of Æthiopia.

The Synonyms, *or Names given to this Tree by different Authors.*

Coffea [*Arabica*], floribus quinquefidis dispermis. Linn. Spec. plant. Ed. II. p. 245.

Jasminum *Arabicum,* lauri folio, cujus semen apud nos Coffé dicitur. Juss. act. Gall. 1713. p. 388, t. 7.

Jas-

Jasminum *Arabicum,* castaneæ folio, flore albo odoratissimo.
Till. Prif. 87. t. 32.
Euonymo similis Ægyptica, fructu baccis lauri simili. Bauh.
Pin. 498. Pluk. Phyt. 272. f. 1.
Bon. Alp. Ægypt. 36. t. 36.

*Explanation of the Letters in the Plate, which have a Reference to
the Diſſection of the Flower and Fruit.*

A. The flower, cut open to shew the situation of the five filaments, with their summits lying upon them.

B. Represents the flower cup, with its four small indentations, inclosing the germen, or embryo seed veſſel; from the middle of which ariſes the style, terminated by the two reflexed spungy tops.

C. The fruit intire; marked at the top with a puncture, like a navel.

D The fruit open, to shew that it conſiſts of two seeds; which are surrounded by the pulp.

E. The fruit cut horizontally, to shew the seeds as they are placed erect, with their flat sides together.

F. One of the seeds taken out, with the membrane or parchment upon it.

G. The same, with the parchment torn open, to give a view of the seed.

H. The seed without the parchment.

THE

HISTORY

OF

COFFEE.

THE earlieft account we have of Coffee is taken from an Arabian manufcript in the king of France's library, N° 944, and is as follows:

Schehabeddin Ben, an Arabian author of the ninth century of the Hegira, or fifteenth of the Chriftians, attributes to Gemaleddin, Mufti of Aden, a city of Arabia Felix, who was nearly his cotemporary, the firft introduction into that country, of drinking Coffee. He tells us, that Gemaleddin, having occafion to travel into Perfia, during his abode there, faw fome of his countrymen drinking Coffee, which at that time he did not much attend to; but, on his return to Aden, finding himfelf indifpofed, and remembering that he had feen his countrymen drinking Coffee in

Perfia,

Perfia, in hopes of receiving fome benefit from it, he determined to try it on himfelf; and, after making the experiment, not only recovered his health, but perceived other ufeful qualities in that liquor; fuch as relieving the head-ach, enlivening the fpirits, and, without prejudice to the conftitution, preventing drowfinefs. This laft quality he refolved to turn to the advantage of his profeffion: he took it himfelf, and recommended it to the Dervifes, or religious Mahometans, to enable them to pafs the night in prayer, and other exercifes of their religion, with greater zeal and attention. The example and authority of the Mufti gave reputation to Coffee. Soon men of letters, and perfons belonging to the law, adopted the ufe of it. Thefe were followed by the tradefmen, and artifans that were under a neceffity of working in the night, and fuch as were obliged to travel late after fun-fet. At length the cuftom became general in Aden; and it was not only drunk in the night by thofe who were defirous of being kept awake, but in the day for the fake of its other agreeable qualities.

The Arabian author adds, that they found themfelves fo well by drinking Coffee, that they entirely left off the ufe of an infufion of an herb, called in their language *Cat*, which poffibly might be Tea, though the Arabian author gives us no particular reafon to think fo.

Before this time Coffee was fcarce known in Perfia, and very little ufed in Arabia, where the tree grew. But, according to Shehabeddin, it had been drunk in Æthiopia from time immemorial.

Coffee, being thus received at Aden, where it has continued in ufe ever fince without interruption, paffed by degrees to many neighbouring towns; and not long after reached Mecca, where

it

it was introduced, as at Aden, by the Dervifes, and for the fame purpofes of religion.

The inhabitants of Mecca were at laft fo fond of this liquor, that, without regarding the intention of the religious, and other ftudious perfons, they at length drank it publicly in coffee-houfes, where they affembled in crouds to pafs the time agreeably, making that the pretence: here they played at chefs, and fuch other kind of games, and that even for money. In thefe houfes they amufed themfelves likewife with finging, dancing, and mufic, contrary to the manners of the rigid Mahometans, which afterwards was the occafion of fome difturbances. From hence the cuftom extended itfelf to many other towns of Arabia, and particularly to Medina, and then to Grand Cairo in Egypt; where the dervifes of the province of Yemen, who lived in a dif-trict by themfelves, drank Coffee the nights they intended to fpend in devotion. They kept it in a large red earthen veffel, and received it refpectfully from the hand of their fuperior, who poured it out into cups for them himfelf. He was foon imitated by many devout people of Cairo, and their example followed by the ftudious; and afterwards by fo many people, that Coffee became as common a drink in that great city, as at Aden, Mecca, and Medina, and other cities of Arabia.

But at length the rigid Mahometans began to difapprove the ufe of Coffee, as occafioning frequent diforders, and too nearly refembling wine in its effects; the drinking of which is contrary to the tenets of their religion. Government was therefore obliged to interfere, and at times reftrain the ufe of it. However, it had become fo univerfally liked, that it was found afterwards neceffary to take off all reftraint for the future.

Coffee continued its progrefs through Syria, and was received at Damafcus and Aleppo without oppofition: and in the year

5　　　　　　　　　　　　　　　　　　　　　1554,

1554, under the reign of the great Soliman, one hundred years after its introduction by the Mufti of Aden, became known to the inhabitants of Conſtantinople ; when two private perſons, whoſe names were Schems and Hekin, the one coming from Damaſcus, and the other from Aleppo, each opened a Coffee-houſe in Conſtantinople, and ſold Coffee publicly, in rooms fitted up in an elegant manner ; which were preſently frequented by men of learning, and particularly poets and other perſons, who came to amuſe themſelves with a game of cheſs, or draughts ; or to make acquaintance, and paſs their time agreeably at a ſmall expence.

Theſe houſes and aſſemblies inſenſibly became ſo much in vogue, that they were frequented by people of all profeſſions, and even by the officers of the ſeraglio, the pachas, and perſons of the firſt rank about the court.　However, when they ſeemed to be the moſt firmly eſtabliſhed, the Imans, or officers of the Moſques, complained loudly of their being deſerted, while the Coffee-houſes were full of company. The Derviſes and the religious orders murmured, and the Preachers declaimed againſt them, aſſerting that it was a leſs ſin to go to a Tavern than to a Coffee-houſe.

After much wrangling, the devotees united their intereſts to obtain an authentic condemnation of Coffee, and determined to preſent to the Mufti a petition for that purpoſe ; in which they advanced, that roaſted Coffee was a kind of coal, and that what had any relation to coal was forbid by law.　They deſired him to determine on this matter according to the duty of his office.

The Chief of the Law, without entering much into the queſtion, gave ſuch a deciſion as they wiſhed for, and pronounced that the drinking of Coffee was contrary to the law of Mahomet.

So

So refpectable is the authority of the Mufti, that nobody dared to find fault with his fentence. Immediately all the Coffee-houfes were fhut, and the officers of the police were commanded to prevent any one from drinking Coffee. However, the habit was become fo ftrong, and the ufe of it fo generally agreeable, that the peoplecontinued, notwithftanding all prohibitions, to drink it in their own houfes. The officers of the police, feeing they could not fupprefs the ufe of it, allowed of the felling it, on paying a tax; and the drinking it, provided it was not done openly; fo that it was drunk in particular places, with the doors fhut, or in the back room of fome of the fhopkeepers houfes.

Under colour of this, Coffee-houfes by little and little were re-eftablifhed; and a new Mufti, lefs fcrupulous and more enlightened than his predeceffor, having declared publicly, that coffee had no relation to coal, and that the infufion of it was not contrary to the law of Mahomet, the number of Coffee-houfes became greater than before. After this declaration, the religious orders, the preachers, the lawyers, and even the Mufti himfelf drank Coffee; and their example was followed univerfally by the court and city.

The Grand Vifirs, having poffeffed themfelves of a fpecial authority over the houfes in which it was permitted to be drunk publickly, took advantage of this opportunity of raifing a confiderable tax on the licences they granted for that purpofe, obliging each mafter of a Coffee-houfe to pay a fequin per day, and limiting however the price at an afper a difh (*a*).

Thus

(*a*) The Turkifh Sequin (according to Chambers) is of the value of about nine fhillings fterling; and the Afper is a very fmall filver coin of the value of fomething more than an Englifh half-penny. The prefent value is neaily feven

fhillings;

Thus far the Arabian manuſcript in the king of France's library, as tranſlated by Mr. Galand; who proceeds to inform us of the occaſion of a total ſuppreſſion of public Coffee-houſes during the war in Candia, when the Ottoman affairs were in a critical ſituation.

The liberty which the politicians who frequented theſe houſes took, in ſpeaking too freely of public affairs, was carried to that length, that the Grand Viſir Kupruli, father of the two famous brothers of the ſame name, who afterwards ſucceeded him, ſuppreſſed them all, during the minority of Mahomet the Fourth, with a diſintereſtedneſs hereditary in his family, without regarding the loſs of ſo conſiderable a revenue, of which he reaped the advantage himſelf. Before he came to that determination, he viſited, incognito, the ſeveral Coffee-houſes, where he obſerved ſenſible grave perſons diſcourſing ſeriouſly of the affairs of the empire, blaming adminiſtration, and deciding with confidence on the moſt important concerns. He had before been in the Taverns, where he only met with gay young fellows, moſtly ſoldiers, who were diverting themſelves with ſinging, or talking of nothing but gallantry and feats of war. Theſe he took no further notice of.

After the ſhutting up of the Coffee-houſes, no leſs Coffee was drunk, for it was carried about in large copper veſſels with fire under them, through the great ſtreets and markets. This was only done at Conſtantinople; for in all other towns of the empire, and even in the ſmalleſt villages, the Coffee-houſes continued open as before.

Notwithſtanding this precaution of ſuppreſſing the public meetings at Coffee-houſes, the conſumption of Coffee increaſed;

ſhillings; that is, two ſhillings and three-pence three-farthings for a dollar, or eighty aſpers; conſequently three aſpers are worth ſomething more than a penny ſterling; but they are generally reckoned at an half-penny each. Two hundred and forty-three aſpers go to a ſequin.

for

for there was no houfe or family, rich or poor, Turk or Jew, Greek or Armenian, who are very numerous in that city, where it was not drunk at leaft twice a day, and many people drank it oftener, for it became a cuftom in every houfe to offer it to all vifitors; and it was reckoned an incivility to refufe it; fo that many people drank twenty difhes a day, and that without any inconvenience, which is fuppofed by this author an extraordinary advantage: and another great ufe of Coffee, according to him, is its uniting men in fociety in ftricter ties of amity than any other liquor; and he obferves, that fuch proteftations of friendfhip as are made at fuch times, are far more to be depended upon than when the mind is intoxicated with inebriating liquors. He computes, that as much is fpent in private families in the article of Coffee at Conftantinople, as in Wine at Paris; and relates, that it is cuftomary there to afk for money to drink Coffee, as in Europe for money to drink your health in Wine or Beer.

Another curious particular we find mentioned here, is, that the refufing to fupply a wife with Coffee, is reckoned among the legal caufes of a divorce.

The Turks drink their Coffee very hot and ftrong, and without fugar. Now and then they put in, when it is boiling, a clove or two bruifed, according to the quantity; or a little of the *femen badian*, called ftarry annifeed, or fome of the leffer cardamums, or a drop of effence of amber.

It is not eafy to determine at what time, or upon what occafion, the ufe of Coffee paffed from Conftantinople to the Weftern parts of Europe. It is however likely that the Venetians, upon account of the proximity of their dominions, and their great trade to the Levant, were the firft acquainted with it; which appears from part of a letter wrote by Peter della Valle, a Venetian, in 1615, from Conftantinople; in which he tells his friend, that, upon his return he fhould bring with him fome Coffee, which he believed was a thing unknown in his country.

C 2

Mr.

Mr. Galand tells us he was informed by Mr. de la Croix, the King's Interpreter, that Mr. Thevenot, who had travelled through the Eaſt, at his return in 1657, brought with him to Paris ſome Coffee for his own uſe, and often treated his friends with it; amongſt which number Monſieur de la Croix was one; that from that time he had continued to drink it, being ſupplied by ſome Armenians who ſettled at Paris, and by degrees brought it into reputation in that city.

It was known ſome years ſooner at Marſeilles; for in 1644, ſome gentlemen who accompanied Monſieur de la Haye to Conſtantinople, brought back with them on their return, not only ſome Coffee, but the proper veſſels and apparatus for making and drinking it, which were particularly magnificent, and very different from what are now uſed amongſt us. However, until the year 1660, Coffee was drunk only by ſuch as had been accuſtomed to it in the Levant, and their friends: but that year ſome bales were imported from Egypt, which gave a great number of perſons an opportunity of trying it, and contributed very much to bringing it into general uſe; and in 1671, certain private perſons at Marſeilles determined for the firſt time to open a Coffee-houſe in the neighbourhood of the Exchange, which ſucceeded extremely well; people met there to ſmoke, talk of buſineſs, and divert themſelves with play: it was ſoon crouded, particularly by the Turkey merchants and traders to the Levant. Theſe places were found very convenient for diſcourſing on and ſettling matters relating to commerce; and ſhortly after, the number of Coffee-houſes encreaſed amazingly. Notwithſtanding which, there was not leſs drunk in private houſes, but a much greater quantity; ſo that it became univerſally in uſe as Marſeilles, and the neighbouring cities.

Before the year 1669, Coffee had not been ſeen at Paris, except at Mr. Thevenot's, and ſome of his friends; nor ſcarce

heard

heard of but from the account of travellers. That year was diftinguifhed by the arrival of Soliman Aga, Ambaffador from Sultan Mahomet the Fourth. This muft be looked upon as the true period of the introduction of Coffee into Paris. For that minifter and his retinue brought a confiderable quantity with them, which they prefented to fo many perfons of the court and city, that many became accuftomed to diink it, with the addition of a little fugar; and fome, who had found benefit by it, did not chufe to be without it. The Ambaffador ftaid at Paris from July 1669 to May 1670, which was a fufficient time to eftablifh the cuftom he had introduced.

Two years afterwards, an Armenian, of the name of Pafcal, fet up a Coffee-houfe, but meeting with little encouragement, left Paris, and came to London; he was fucceeded by other Armenians and Perfians, but not with much fuccefs, for want of addrefs and proper places to difpofe of it; genteel people not caring to be feen in thofe places where it was fold. However, not long after, when fome Frenchmen had fitted up for the purpofe fpacious apartments in an elegant manner, ornamented with tapeftry, large looking-glaffes, pictures, and magnificent luftres, and began to fell Coffee, with Tea, Chocolate, and other refrefhments, they foon became frequented by people of fafhion and men of letters, fo that in a fhort time the number in Paris increafed to three hundred.

For this account of the introduction of the ufe of Coffee into Paris, we are indebted to La Roque's Voyage into Arabia Felix. We now come to trace its firft appearance in London.

It appears from Anderfon's Chronological Hiftory of Commerce, that the ufe of Coffee was introduced into London fome years earlier than into Paris. For in 1652 one Mr. Edwards, a Turkey merchant, brought home with him a Greek fervant, whofe name was Pafqua, who underftood the roafting and

making

making of Coffee, till then unknown in England. This servant was the first who sold Coffee, and kept a house for that purpose in George-yard, Lombard-Street.

The first mention of Coffee in our statute books, is anno 1660 (12 Car. II. cap. 24.) when a duty of Four-pence was laid upon every gallon of Coffee made and sold, to be paid by the maker.

The statute of the 15 Car. II. cap. xi. § 15. ann. 1663 directs that all Coffee-houses should be licensed at the General Quarter Sessions of the Peace for the county within which they are to be kept.

In 1675, King Charles issued a Proclamation, to shut up the Coffee-Houses, but in a few days suspended the proclamation by a second. They were charged with being seminaries of sedition.

The first European author who has made any mention of Coffee, is Rauwolfus, who was in the Levant in 1573; but the first who has particularly described it, is Prosper Alpinus, in his History of the Egyptian Plants, published at Venice in 1591, whose description we have in Parkinson's Theatre of Plants, page 1622, chap. 79. as follows:

Arbor Bon, cum fructu suo buna, the Turk's berry drink; Alpinus in his book of Egyptian Plants, gives us the description of this tree, which he says, he saw in the garden of a captain of the Janissaries, which was brought out of Arabia Felix and there planted, as a rarity never seen growing in those places before. The tree, saith Alpinus, is somewhat like the Euonymus, or Spindle tree, but the leaves of it were thicker, harder, and greener, and always abiding on the tree. The fruit is called *Buna,* and is somewhat bigger than a hazel nut, and longer, round also, and pointed at one end, furrowed likewise on both sides, yet on one side more conspicuous than the other,

that

that it might be parted into two, in each fide whereof lieth a fmall oblong white kernel, flat on that fide they join together, covered with a yellowifh fkin, of an acid tafte, and fomewhat bitter, and contained in a thin fhell (*b*) of a darkifh afh colour. With thefe berries, in Arabia and Ægypt, and other parts of the Turkifh dominions, they generally make a decoction or drink, which is in the ftead of wine to them, and commonly fold in their Tap-houfes, or Taverns, called by the name of *Caova*. Paludamus fays, *Choava*, and Rauwolfus, *Chauke*. This drink has many good phyfical properties : it ftrengthens a weak ftomach, helping digeftion, and the tumours and obftructions of the liver, and fpleen being drank fafting for fome time together. It is held in great eftimation among the Ægyptian and Arabian women, in common feminine cafes, in which they find it does them eminent fervice.

Lord Chancellor Bacon likewife makes mention of it in 1624; he fays, that the Turks have a drink called Coffee, made with boiling water, of a berry reduced into powder, which makes the water as black as foot, and is of a pungent and aromatic fmell, and is drunk warm.

The celebrated John Ray, in his Hiftory of Plants, publifhed in 1690, fpeaking of it as a drink very much in ufe, fays, that this tree grows only within the Tropics, and fuppofes that the Arabs deftroy the vegetable quality of the feeds, in order to confine among themfelves the great fhare of wealth, which is brought thither from the whole world for this commodity : from whence he obferves that this part of Arabia might be truly ftyled the moft happy, and that it was almoft incredible how many millions of bufhels were exported from thence into Turkey, Barbary, and Europe ; he fays, he was aftonifhed that

(*b*) This defcription is evidently taken from a dried berry, and not from the ripe fruit.

one

one particular nation fhould poffefs fo great a treafure; and that within the narrow limits of one province; and that he wondered the neighbouring nations did not contrive to bring away fome of the found feeds or living plants, in order to fhare in the advantages of fo lucrative a trade.

We now come to fhew by what means this valuable tree was firft introduced into Europe, and from thence into America.

The firft account of this tree being brought into Europe, we have from Boerhaave, in his Index of the Leyden Garden, part II. page 217, which is as follows: Nicholas Witfen, burgomafter of Amfterdam, and governor of the Eaft India Company, by his letters often advifed and defired Van Hoorn, governor of Batavia, to procure from Mocha, in Arabia Felix, fome berries of the Coffee-tree, to be fown at Batavia; which he having accordingly done, and by that means, about the year 1690, raifed many plants from feeds, he fent one over to governor Witfen, who immediately prefented it to the garden at Amfterdam, of which he was the founder and fupporter: it there bore fruit, which in a fhort time produced many young plants from the feeds. Boerhaave then concludes that the merit of introducing this rare tree into Europe, is due to the care and liberality of Witfen alone.

In the year 1714, the magiftrates of Amfterdam, in order to pay a particular compliment it Lewis XIV, king of France, prefented to him an elegant plant of this rare tree, carefully and judioufly packed up to go by water, and defended from the weather by a curious machine covered with glafs. The plant was about five feet high, and an inch in diameter in the ftem, and was in full foliage, with both green and ripe fruit. It was viewed in the river, with great attention and curiofity, by feveral members of the Academy of Sciences, and was afterwards conducted to the Royal Garden at Marly, under the care of

Monfieur

Monfieur de Juffieu, the king's profeffor of Botany; who had, the year before, written a Memoir, printed in the Hiftory of the Academy of Sciences of Paris, in the year 1713, defcribing the charaƈters of this genus, together with an elegant figure of it, taken from a fmaller plant, which he had received that year from Monfieur Pancras, burgomafter of Amfterdam, and direƈtor of the botanical garden there.

In 1718, the Dutch colony at Surinam began firft to plant Coffee; and in 1722, Monfieur de la Motte Aigron, governor of Cayenne, having bufinefs at Surinam, contrived, by an artifice, to bring away a plant from thence, which, in the year 1725, had produced many thoufands.

In 1727, the French, perceiving that this acquifition might be of great advantage in their other colonies, conveyed to Martinico fome of the plants; from whence it moft probably fpread to the neighbouring iflands: for in the year 1732, it was cultivated in Jamaica, and an aƈt paffed to encourage its growth in that ifland.—Thus was laid the foundation of a moft extenfive and beneficial trade to the European fettlements in the Weft-Indies.

An

An Account of the Culture of the Coffee Tree in Arabia Felix, extracted from La Roque's Voyage.

HE relates, that the Coffee-tree is there raised from feed, which they fow in nurferies, and plant them out as they have occafion. They chufe for their plantations a moift, fhady fituation, on a fmall eminence, or at the foot of the mountains; and take great care to conduct from the mountains little rills of water, in fmall gutters or channels, to the roots of the trees; for it is abfolutely neceffary they fhould be conftantly watered, in order to produce and ripen the fruit. For that purpofe, when they remove or tranfplant the tree, they make a trench of three feet wide, and five feet deep, which they line or cover with ftones, that the water may the more readily fink deep into the earth, with which the trench is filled, in order to preferve the moifture from evaporating. When they obferve that there is a good deal of fruit upon the tree, and that it is nearly ripe, they turn off the water from the roots, to leffen that fucculency in the fruit, which too much moifture would occafion.

In places much expofed to the South, they plant their Coffee-trees in regular lines, fheltered by a kind of Poplar-tree, which extends its branches on every fide to a great diftance, and affords a very thick fhade. Without fuch precaution they fuppofe the exceffive heat of the fun would parch and dry the bloffoms fo, that they would not be fucceeded by any fruit.

In fituations not fo much expofed to the fun, this defence is not neceffary. When they perceive the fruit come to maturity, they

fpread

spread cloths under the trees, which they shake, and the ripe fruit drops readily [c]; they afterwards spread the berries upon mats, and expose them to the sun, until they are perfectly dry. After which they break the husk with large heavy rollers, made either of wood or stone. When the Coffee is thus cleared of its husk, it is again dried in the sun; for, unless it is thoroughly dried, there is danger of its heating on board the ship. It is then winnowed with a large fan; for if it is not well cleaned and dried, it sells for a much lower price.

The manner of preparing and drinking Coffee *among the* Arabians, *from the same Author.*

THE Arabians, when they take their Coffee off the fire, immediately wrap the vessel in a wet cloth, which fines the liquor instantly, makes it cream at top, and occasions a more pungent steam, which they take great pleasure in snuffing up as the Coffee is pouring into the cups. They, like all other nations of the East, drink their Coffee without sugar.

People of the first fashion use nothing but Sultana Coffee, which is prepared in the following manner: they bruise the outward husk, or dried pulp, and put it into an iron or earthen pan, which is placed upon a charcoal fire; they then keep stirring it to and fro until it becomes a little brown, but not of so deep a colour as common Coffee; they then throw it into boiling water, ad-

[c] This circumstance deserves the particular attention of the West India planter, who, I am told, is accustomed to gather his Coffee as soon as it turns red, before it changes to a dark red colour, and begins to shrivel; whereas the Arabians wait for those tokens, which shew the full maturity of the fruit. Mr. Miller in his Dictionary mentions, that in some stoves in England, Coffee is raised of a better quality than the best Mocha Coffee that can be procured in this country; which may likewise be owing to gathering the fruit only when it is thoroughly ripe.

ding

ding at leaſt the fourth part of the inward huſks; which is then boiled all together in the manner of other Coffee: the colour of this liquor has ſome reſemblance to the beſt Engliſh beer. The huſks muſt be kept in a very dry place, and packed up very cloſe; for the leaſt humidity ſpoils the flavour. They eſteem the liquor prepared in this manner preferable to any other. The French, when they were at the court of the King of Yemen, ſaw no other Coffee drunk, and they found the flavour of it very delicate and agreeable; there was no occaſion to uſe ſugar, as it had no bitter taſte to correct. In all probability, this Sultana Coffee can only be made where the tree grows; for as the huſks have little ſubſtance, if they are too much dried in order to ſend them to other countries, the agreeable flavour they had when freſh is greatly impaired.

It may perhaps be worth while for our Weſt India planters, to make a trial of drying the outward and inner huſk of Coffee, ſeparately, in the manner the Chineſe do their Tea, upon a broad, ſhallow iron pan, turned upwards at the brim, placed upon a ſtove. They ſhould be kept continually turning, to prevent burning; and when they are become too hot to be handled, they ſhould be taken off with a kind of ſhovel, and laid upon a matt, on a low table, and ſhifted about until they cool, fanning them at the ſame time, to diſperſe the moiſture. The pan muſt be frequently wiped and kept clean from any clammy matter ſticking to it, and the proceſs repeated while any moiſture is perceived. They muſt afterwards be packt cloſe in dry jars, canniſters, or cheſts, lined with lead, ſuch as the Tea is ſent over in. It will be proper to turn out theſe huſks, after they have lain ſome days, to examine whether they are thoroughly dry; and if the leaſt damp is felt, it will be neceſſary to dry them ſtill more, otherwiſe they will become mouldy, and loſe their flavour. For it appears

pears from the Arabian account, that they are not acquainted with a proper method of drying thefe hufks, and packing them fo as to be conveyed to any confiderable diftance, without pre-judicing this agreeable flavour.

The Chinefe are very careful not to leave their Tea leaves in heaps before they are dried; which would occafion them to heat and fpoil. They likewife gather no more at a time than they can dry in lefs than 24 hours; as they find, when they have been kept longer, they turn black. Thefe obfervations may poffibly be of fome ufe to thofe who may be induced to attempt drying the pulp of the berry, for the purpofe of making Sultana **Coffee.**

Extract

Extract from Nieburh's *Voyage to* Arabia, *lately published in* Denmark.

THE Arabians drink but little with their meals, but soon after them take a good draught of water, and thereupon a cup of Coffee, without milk or sugar; but prepared in other respects in the same manner as ours. However, this liquor is rarely drunk in Yemen, because it is there believed to heat the blood. But the inhabitants of that province compose a drink of the hulls of Coffee, which in taste and colour much resembles Tea: this they esteem wholesome and refreshing. It is prepared nearly in the same manner as that from the seed or bean, and is the " Caffée à la Sultane" of the French.

Nieburh's party had taken with them a Coffee-mill to Arabia, but soon left off using it, because they found the ground Coffee much inferior to the bruised; which last is the Arabian method of preparing it.

The Coffee trees are particularly cultivated to the West of the great mountains which run through Yemen. The exportation of this plant is forbidden, under the severest penalties; and yet the Dutch, French, and English, have found means to transport some of them into their colonies; but the Coffee of Yemen still keeps the preference, probably because the Europeans do not cultivate theirs in the same manner, and upon such high mountains, where there is so regular a temperature of air as in Yemen.

The English East India Company send only one vessel every second year into the Arabian gulph, to take in there a lading of Coffee.

O B.

OBSERVATIONS ON COFFEE,

From Dr. BROWN's Natural Hiſtory of Jamaica, p. 161.

THIS ſhrub has been long introduced and cultivated in the iſland of Jamaica, where it grows very luxuriantly, and riſes frequently to the height of eight or nine feet, ſpreading its flexile branches to a conſiderable diſtance on every ſide. It thrives beſt in a rich ſoil, and cool, ſhaded ſituation, where it can be duly refreſhed with a moderate ſhare of moiſture : and in ſuch a ſoil and ſituation, it generally produces ſo great a quantity of fruit, that the branches can hardly ſuſtain the weight, though bending to the ground; and you may frequently obſerve the very trunk to yield to the load. The tree, however, is obſerved to grow and thrive in almoſt every ſoil about the mountains, and will frequently produce great quantities of fruit in the dryeſt ſpots; though in Arabia, where this plant is a native, and had been firſt propagated, and brought into uſe, it is obſerved to be cultivated between the hills : and yet the drought of the place is ſuch, that they are frequently obliged to refreſh the roots with water; which, as it is often wanted in that country, is generally conveyed by gutters or channels through every piece.

It is a general remark in England, and indeed a certain one, that the Coffee imported from America does not anſwer ſo well as that of the growth of Arabia; nor is it owing, as ſome imagine, to any foreign fume, or vapours it might have contracted in the paſſage, though great care ſhould be always taken to prevent

any

any acquifition of this nature; for even there what is commonly ufed, will neither parch or mix like the Turkey Coffee: but this has been hitherto owing to the want of obfervation, or knowing the nature of the grain; moft people being attentive to the quantity of the produce, while the qualities are but feldom confidered.

I have been many years in thofe colonies; and, being always a lover of Coffee, have been often obliged to put up with the produce of the country in its different ftates. This gave me room to make many obfervations upon this grain; and I dare fay they are fuch as will be conftantly found true; and, if rightly regarded, will foon put the inhabitants of our American colonies in a way of fupplying the mother country with as good Coffee as we ever had from Turkey, or any other part of the world. For the eafier underftanding of this affertion, I fhall fet down the remarks I have made, as they occur.

1. New Coffee will never parch, or mix well, ufe what art you will. This proceeds from the natural clamminefs of the juices of the grain, which requires a fpace of time proportioned to its quantity to be wholly deftroyed.

2. The fmaller the grain, and the lefs pulp the berry has, the better the Coffee, and the fooner it will parch, mix, and acquire a flavour.

3. The drier the foil, and the warmer the fituation, the better the Coffee it produces will be, and the fooner it will acquire a flavour.

4. The larger and the more fucculent the grain, the worfe it will be, the more clammy, and the longer in acquiring a flavour.

5. The worft Coffee produced in America will, in a courfe of years, not exceeding ten or fourteen, be as good, parch, and

mix

mix as well, and have as high a flavour, as the beſt we now have from Turkey; but due care ſhould be taken to keep it in a dry place, and to preſerve it properly.

6. Small grained Coffee, or that which is produced in a dry ſoil and warm ſituation, will in about three years be as good, and parch as well, as that which is now commonly uſed in the Coffee-houſes in London.

Theſe are facts founded on repeated experiments, which I have tried from time to time, during my reſidence in Jamaica; though it be very rare to ſee what a man may call good Coffee in the iſland, for they generally drink it à la Sultane (*d*), and never reſerve more than is ſufficient to ſupply them from one year to another.

I have examinéd the Turkey Coffee with great care ſince I came to England; and conclude, from the ſize of the grain, the frequent abortion of one of the ſeeds, and the narrowneſs of the ſkin·that contains the pulp, that the ſhrub muſt be greatly ſtunted in its growth; and from hence judge, that whoever endeavours to produce good Coffee, and ſuch as would mellow as ſoon as that of Arabia, or expect ſeeds that may have the ſame flavour, muſt try what can be produced in the lower hills and mountains of the Southern part of the iſland: nay, even try what the ſavannas will bear; and I am perſuaded it would anſwer well in many places about the foot of the long mountain near Kingſton: an acre or two may be eaſily tried in any part, and the experiment will be well worth the labour; but who-

(*d*) This I take to be rather the infuſion of the half burnt flakes of new Coffee, (for it never will parch, or mix properly while freſh.) like that commonly uſed by the Coffee planters in Jamaica, than a decoction of the coverings, as it is commonly reported to be.—This appears to be a miſtake of Dr. Brown's, when we conſider the account given by the French, who travelled in Arabia Felix,,who have very fully deſcribed the manner of making the Sultana Coffee.

E   ever

ever is for having greater crops, muſt keep among the mountains, where the trees grow and ſhoot out more luxuriantly.

Wherever this plant is cultivated, it ſhould be planted at
diſtances proportioned to its growth; for in a dry, gravelly, or
mixed ſoil, it ſeldom riſes above five feet, and may be conve-
niently planted within that diſtance of each other. But among
the mountains of Jamaica, where it frequently riſes to the
height of nine or ten feet, or more, it requires a larger ſcope,
and in ſuch a ſoil can be hardly planted nearer than eight or ten
feet from each other: I have, however, frequently known
them crowded in ſuch places, and yet produce a great quantity
of fruit.

The gentlemen of Jamaica imagine, that a great deal of the
richneſs and flavour of the Turkey Coffee, depends upon their
methods of drying it : but this is an ill-grounded notion, for the
berries, as well as the trees, being naturally ſtunted in their
growth in moſt parts of Arabia, they have but little pulp, and
are very eaſily dried in that warm climate, where a few days
ſun generally compleats the work, without being at the trouble
of ſtripping them of any part of their more juicy coats before-
hand : but though I am ſatisfied the Turkey Coffee receives no
addition from any peculiar method of drying it; I am equally
convinced that great quantities of that produced in the woody
parts of Jamaica, where the berries are large and ſucculent,
and the ſeeds lax and clammy, are greatly prejudiced by the
methods uſed there : ſuch berries ſhould be undoubtedly ſtripped
of a great part of the pulp, and the ſeeds carried down to the low
lands to be dried ; and not left ſoaking in their clammy juices,
to dry but ſlowly in a damp air, as they generally do in many
parts of that iſland : but this is no prejudice to the ſale of it
among

among the Northern purchasers, who generally look upon the largest and fattest grain as the best ; nor do they chuse it by any other marks than the plumpness of the seeds, and a fresh colour, which generally is a blueish pale in new Coffee.

Such as have large Coffee-walks, should be provided with a convenient *barbakue*, or platform, to dry these seeds more commodiously upon ; and I think it would be well worth while to try whether sweating would destroy any of the clamminess peculiar to the seeds of the larger berries ; but these should be always pulped and dried as soon as possible ; nor do I imagine but the ease and speed whereby they might be dried in the low lands, would be a sufficient recompence for the trouble of carrying them there as they are picked from the trees.

After the fruit is well dried, it must be husked, and the seeds cleared from all the outward coverings, to fit and prepare them for the market. This is generally done in Jamaica, by pounding the dried berries lightly in large wooden mortars, until, after a long continued labour, both the dried pulp, and inward membraneous coverings, are broke and fall to pieces among the seeds. The whole is then winnowed, cleared, exposed afresh to the sun for some days, and then casked for the market. But the Arabians, after having dried their Coffee sufficiently on matts, spread it on an even floor, and break off the covering, by passing a large weighty roller, of some heavy wood or stone, to and fro upon it : and when the husks are well broke in this manner, it is winnowed, and exposed to the sun anew, until it is very well dried ; for otherwise it is apt to heat on board ships, and then it loses all its flavour.

The drink prepared from the seed of this plant, is now generally used all over Europe, and many parts of Asia and America ; it is generally esteemed as an excellent stomachic and strengthener

of

of the nerves; and peculiarly adapted for ſtudious and ſedentary people.

The plants are propagated by the ſeeds; and, to raiſe them ſuccefsfully, the whole berries ſhould be ſown ſoon after they are gathered from the trees; for if they be kept but a ſhort time out of ground, they are apt to fail; but when the plants riſe about five or ſix inches above the earth, if double (as they generally are) they ſhould be ſeparated, which is done by drawing one or both, parting the roots, and planting them again in ſeparate beds. When the young plants are removed from a bed, or from under the parent-tree, where they generally grow in great abundance, great care ſhould be taken not to break or injure the roots, and to preſerve the earth about them until they are replanted; for if the fibres are expoſed to the air, and allowed to dry, they are very ſubjeƈt to periſh; which is the reaſon they have not this beautiful tree more common in the gardens about the lower lands of Jamaica, where very few tranſplants of the kind thrive, being generally pulled up very bare, the layers laid by commonly for thirty or forty hours afterwards, and then carried a conſiderable diſtance in the heat of the ſun; but ſuch as would have them proſper well, ſhould be careful to procure plants that are well ſupplied with mould from their native beds; or to raiſe them immediately from the ſeeds.

Extraƈt

ExtraEt of a Letter from Dr. Fothergill *to* J. Ellis, *Efquire,* F. R. S. *Agent for* Dominica. *Containing fome Remarks on the Culture and Ufe of Coffee.*

Sept. 2. 1773.

IT is doing a very ufeful piece of fervice, and I believe, an acceptable one to the publick, to make them a little better acquainted with Coffee, which now conftitutes fo confiderable a part of their entertainment, if not fuftenance. I am pleafed with the engraving of this very elegant plant; it is executed in the beft manner of this able artift (*e*), and exactly after the finifhed drawing he received.

I have not time to collect, or relate with fufficient accuracy, the hiftory of this berry, fo far as it might be traced in the Afiatic hiftories: It has been ufed for ages. By the account which is fubjoined to the reflections I am going to make on this fubftance, it will appear, that it was introduced by the French into Martinico in the Weft Indies no longer ago than the year 1727; that it has been fince that time propagated in almoft all the Weft India iflands, Englifh, French, and Dutch; though it has not been cultivated by us with the attention it deferves.

The greateft part of the Coffee now made ufe of in Europe is, I believe, the produce of the Weft Indies; at leaft, the confumption of Mocha Coffee amongft us feems to be greatly reduced.—Several years ago two fhips were fent out annually by the Eaft India Company; they now only fend one every two years for this article, if I am rightly informed; though at the fame time it may be prefumed a much larger quantity is confumed than at any time heretofore.

(*c*) J. S. Miller.

2

The

The French, and other nations, who have poſſeſſions in the Weſt Indies, ſupply us clandeſtinely with large quantities. It is true, we import a great deal of Raw Coffee from our own iſlands; but the beſt is of foreign growth. The French in particular cultivate it with great attention; much likewiſe is brought from their Eaſt India ſettlements. Thoſe who are accuſtomed to drink Coffee frequently, are ſenſible of a very manifeſt difference between the Aſiatic, the French, and American Coffee. The refreſhing odour of the firſt, and its grateful taſte, much ſurpaſs the beſt Weſt India Coffee I have ever ſeen imported. There is ſomething in the ſmell, a rank-neſs in the taſte, and diſguſting return, eſpecially of that from the Engliſh iſlands, which makes it very unpleaſant to thoſe who have been accuſtomed to the beſt Mocha Coffee.

The tree that was firſt carried to Martinico, was a deſcend-ant of one from Batavia. The Dutch moſt probably brought the plants to their colony from Mocha, and there ſeems no doubt but it is the true Arabian Coffee which is now cultivated in the Weſt Indies. But if we reflect upon the courſe it has taken, we may perhaps ſee cauſe to apprehend that it may have degenerated conſiderably.

That part of Arabia from whence the Aſiatic Coffee is brought, is for the moſt part extremely ſandy, dry, and hot.

At Batavia the ſoil is in general rich and deep; and though, like other eaſtern climates, there is a dry ſeaſon; yet in the rainy periods the quantity of wet that falls is exceſſive. The rich luxuriant ſtate of vegetation in the iſland of Java, on which Batavia is ſituated, is a proof of this aſſertion; and one may ſafely infer, that a plant brought from a dry, ſterile, ſandy ſoil, will aſſume not only a very different appearance, but its fruit will have a very different quality from that which is the

5   produce

produce of a fertile, moiſt, ſoil, ſubjected to equal heat. It is not therefore improbable, but from this circumſtance the plant brought from Batavia to the royal garden in France, and its iſſue tranſported to a climate much more abounding with moiſture than that of which it was a native, may ſo far have aſſumed another nature, as not eaſily to be brought back to its original excellence.

I wiſh this circumſtance however only to be conſidered as a ſuggeſtion, which, though not without the appearance of probability, may not be ſufficiently warranted by experience. But ſhould it prove true, it may lead us to one practicable method of meliorating Coffee. Let the Coffee be planted in a ſoil as ſimilar to its natural one as poſſible. Indeed the ſhort account which is annexed to this letter, confirms my apprehenſions. The dryer the ſoil on which the Coffee grows, the ſmaller is its fruit, and its quality more excellent. There are ſome kinds of trees, perhaps the greateſt part, whoſe fruit, while the trees are young, is either more inſipid, or the taſte of it leſs refined, than at a more advanced age. The fruit of young walnut trees is large; but it is watery and inſipid; as the tree grows older, the nuts decreaſe in ſize, but their taſte is more agreeable. A ſimilar progreſs may be obſerved in many other ſpecies; and it is not improbable but he Coffee-tree may be another inſtance of the like properties. It is certain, that in old Coffee-trees the fruit is ſmaller; perhaps an accurate taſte would diſcover that its flavour is improved in proportion. The experiment may be recommended to thoſe who cultivate the Coffee-tree in our iſlands. But I have not time to trace all the circumſtances that have a probable tendency to leſſen the value of our own plantation Coffee.

I haſten

I haften to another point, which would foon put our planters upon overcoming every difficulty, and would oblige them to ftudy the culture of the plant, the curing of the fruit, and fending it to us in the higheft perfection poffible. By what means can we make it the Weft India planters intereft to cultivate Coffee in fuch a manner, as to approach in tafte and flavour as near to the Afiatic as poffible? Perhaps the fhorteft anfwer to this would be, make it their Intereft; that is, to encourage its importation.

I am well informed, by a perfon intelligent in thefe matters, that the duties and excife on Coffee from the plantations, are as follows:

	£.	s.	d.
The duty on Coffee of the growth of the Britifh plantations, for home confumption, is 1l. 13s.6$\frac{3}{20}$d. per hundred weight; which is per lb. about	0	0	4
Excife on ditto, is, per lb. — —	0	1	6
Total per lb. is	0	1	10

When fuch an exceffive load of expences, and fo many difficulties arife to the grower, importer, and of confequence to the confumer of Weft India Coffee, it is no wonder that the planters give themfelves very little concern about its cultivation. At prefent there is very little difference in the produce, and confequently in the price; the high duties are a bar to its ufe amongft us, the Coffee is in general bad, and the price in proportion.

This

This difcouragement renders them lefs folicitous about it: bad as it may be produced, it finds confumers abroad, and to vend it with certainty, anfwers their purpofe better than a more attentive cultivation of a commodity clogged with fuch duties.

Thofe who know the tafte of Mocha Coffee, and are defirous of ufing our Weft Indian, foon quit it with difguft. Better Coffee than our own, the produce of the French Ifles, finds its way into fome of the out-ports clandeftinely; is much ufed, and thought to be equal to the Turkey. Tafte is perhaps more the effect of habit than is generally admitted; of this, tobacco is the ftrongeft and firft inftance that occurs to me: To a perfon unaccuftomed to it, the fineft is fcarcely tolerable.

If the duties and excife upon Coffee were leffened, the confumption would be encreafed. Tafte would grow more refined; the beft would be fought for, and the price would be in proportion; the prefent duties are almoft prohibitory. It may be worth one's while to view the effects of thefe high duties in a political light—I mean in refpect to this article.

For a century to come, it is perhaps more than probable, that the people of this country will, for one meal at leaft, make ufe of either Tea, Coffee, or Chocolate; I fpeak of the generality. Tea, at prefent, takes the lead: whence it comes, its hiftory, properties, and ufes, have been fo fully explained, that I fhall fay nothing here upon the fubject (*f*).

It is a queftion often propofed to phyficians, which is beft, Tea, or Coffee? The folution of this point would perhaps be a difficult one. We neither find the Chinefe or Turks fubjected

(*f*) See Dr. Letfom on the Tea Plant.

F   to

to any fuch difcriminating effects, as enable the faculty to fay, with precifion, that one is more injurious than the other. For my own part, I leave it to the experience of individuals. To fome people Coffee is difagreeable; they charge it with producing nervous complaints. Tea is not without fimilar accufations. It feems as if the human frame was, however, fo happily conftructed, that it is lefs in the power of fuch things to affect it, than might at firft fight be imagined. The animal powers are apparently fuch, as can convert almoft oppofite principles to its benefit, if ufed in any degree of moderation: fome drink Coffee almoft to excefs, and condemn Tea as injurious; and fo Coffee is treated in its turn. Thefe are proofs, however, how few people are capable of making proper inferences from experiment.

I think neither Coffee nor Tea afford any very material fupport; that is, contain very little nutriment: they are rather the vehicles of nourifhment, than nutritious of themfelves: the moft that can be expected from them in general is, that they are grateful, and very little injurious. Cuftom has adopted them both; and it becomes us to make them as ufeful to ourfelves, and as fubfervient to public good, as may be in our power. China, that fupplies us with Tea, is remote; the navigation long and dangerous; the climate not always favourable to our feamen; indeed, all long voyages are injurious, and the hotter the climate the worfe. As a nation, a commercial nation, whofe accommodations depend on this ufeful race of people, we cannot, as friends to humanity, wifh to promote the confumption of thofe articles, which are introduced at fo great an expence of ufeful lives. Coffee from our own plantations is in this refpect much preferable to Tea; the voyage is fhorter, the

rifque

rifque is lefs. Suppofing then, that Tea and Coffee are alike, in refpect to real ufefulnefs; that one is not inferior to the other in refpect to the health of the confumers: fuppofe, likewife, that the difadvantage with refpect to the lives of the feamen were equal, which however is not the cafe, there is one material difference that ought to turn the fcale in favour of the more general ufe of Coffee. It is raifed byour fellow-fubjects, and paid for with our manufactures. Tea, on the contrary, is paid for principally with money. The quantities of Britifh goods which the Chinefe take from us is inconfiderable, when compared with the quantities we pay for in bullion.

The Chinefe take from us every article which they can turn to national benefit; and whatever enables them to improve their manufactures. Befides Raw Silk, and a few other articles of fome little ufe in our own manufactures, moft other things imported from thence we can do without, efpecially if the confumption of our Coffee was encouraged. Were the duties and excife upon Coffee, for inftance, reduced to a quarter part, more than double the quantity would be confumed. Was the confumption greater, the planters would find it their intereft to cultivate the trees with more attention. Increafed demand would increafe the price; and as more came to market, the beft would fell dearer than an inferior kind. Thefe muft be the certain effects of increafed demand.

There is another confideration of fome moment likewife; which is, that the cultivation of Coffee might be carried on in fuch manner, as the leffer planters might fubfift by it; and a few

fimilar

fimilar articles, cotton particularly, with little ftock, and without much expence for Negroes. No little planter can make fugar to advantage. The expence of Negroes, cattle, mills, and other requifites of a fugar plantation, are beyond his reach. If he has any landed property, by one means or another, he is often obliged to fell it to his richer neighbour; and to re-move to fome other country, lefs unfavourable to contracted circumftances. Thus the iflands are gradually thinned of the white inhabitants; they become lefs able to quell the infurrections of their Negroes, or to oppofe any hoftile invafion.

The annexed account of Coffee anticipates fome remarks, I meaned to have fpoke to more fully, which had often occurred to me. The writer of that fhort account has not, however, wholly exhaufted the fubject. He very juftly defcribes many circumftances which tend to make Weft India Coffee of lefs value than the European. He is very right in his obferva-tions on the difference of quantity produced in different foils and fituations. He moft pertinently cenfures the Englifh for want of care in fhipping it home. The French exceed us vaftly in this refpect; and the greater price it fetches, is owing in a great degree to fuperior care and management. One would hardly fuf-pect the merchants and planters could be capable of fo much inattention as to fhip Coffee in veffels loaded with Rum and coarfe Sugars; articles capable of communicating a tafte fcarce to be driven off by fire: fo penetrating are the fteams arifing from Rum and Sugars confined in a fhip's hold. So much Coffee ought to be collected together at one place, as to load a veffel. It is objected likewife, that the Coffee in the Weft India iflands

cannot

cannot eafily be dried in a proper manner, from the great moifture of the air. But there are in all the iflands high grounds, to which the Coffee might be brought and dried fufficiently.

Another point ought not to be omitted, which is, that our plantation Coffee is made ufe of too foon. Perhaps one part of the excellence of Mocha Coffee arifes from this circumftance—The Eaft India Company fend a fhip once in two years: it is moft probable a part of the loading has been kept in that hot, dry country above a year: it is fix months before it arrives in England; it may be fix or twelve months more before it comes into the confumer's hands. Thus, between two and three years muft inevitably intervene between its growth and confumption.

Much of that mucilage, which moft probably in roafting is the bafis of its flavour, is changed by this delay; and indeed experience confirms it.

Befides many inftances that might be given from credible witneffes, (and efpecially from governor Scott's account of this fubject, hereunto annexed) the following paffed under my own obfervation, and, as far as it reaches, may be conclufive.

I had a prefent made of feveral kinds of raw Coffee from the Weft India iflands; it being known that I wifhed to encourage the culture of this plant, for the reafons I have alledged. Some of this Coffee, which a year ago was fo ill tafted as to be unfit for ufe, was laid in a very dry clofet: this year it was again tried, and found to be greatly amended; in another, it will probably be little inferior to the Afiatick, if it amends in proportion. It is of much confequence whether the Coffee is imported with other goods, or alone; whether it is kept in moift, damp warchoufes, or in dry, airy places; whether it is ufed immediately, or not till after it has been kept a confiderable time. It would be well worth the planter's labour and expence, to keep his Coffee in the

ifland.

island from year to year, till he has got such a quantity, either of his own, or bought from his neighbours, sufficient to load a small veffel; marking the different ages. But the whole of this depends entirely on government. Leffening the duty would encreafe the confumption; prevent fmuggling; enable many whites to gain a comfortable fupport, and to pay for our manufactures. As it is raifed by our own people, imported with lefs rifque of feamens health and lives, in a political light, it muft certainly deferve the deliberate attention of the legiflature.

Coffee made in the following manner is pleafing to moft people, and is much preferable to Tea, or to Coffee made in the ufual manner, for breakfaft. Let Coffee be made in the ufual manner, only a third part ftronger; let as much boiling milk be added to the Coffee before it is taken from the fire, as there is water; let it fettle; drink it with cream, or without, as may be moft agreeable. And were the poor and middling people enabled to procure this, it would be much more nourifhing and beneficial, than the wretched beverage they indulge themfelves with of the moft ordinary Teas. Very little fugar ought to be ufed with Coffee; on weak ftomachs it is too apt to become acid, if made fweet: and this is one reafon why many people forbear drinking Coffee. I do not prefume to fettle this important queftion, which is preferable, Tea or Coffee? This muft be left to the experience of individuals. So far as concerns myfelf, I may be permitted to become evidence.

Though I like Tea, I found it not quite favourable to my health, from fome circumftances. I tried Coffee, made in the manner above-mentioned, and have drunk it almoft conftantly many years, without receiving any inconvenience from it.

It

It may require a good deal of phyſical ſagacity to determine how far the French cuſtom of drinking Coffee immediately after dinner is right ; but I think it can admit of no diſpute whether a diſh of Coffee or a bottle of wine may then be leſs prejudicial to health.

I think however it is leſs injurious to drink Coffee immediately after dinner, than later in the evening ; and at leaſt for one very obvious reaſon.

Coffee moſt certainly promotes watchfulneſs ; or, in other words, it ſuſpends the inclination to ſleep. To thoſe therefore who wiſh not to be too ſubject to this inclination, Coffee is undoubtedly preferable to wine, or perhaps to any other liquor we know.

The inſtances of perſons to whom Coffee has this antiſoporific effect are very numerous. And the inſtances are almoſt as numerous of ſuch to whom wine has the oppoſite effect.

To attribute the livelineſs of the French, after their repaſts, to this beverage, would be highly hypothetical. But I think it muſt be acknowledged that, after a full meal, perhaps of groſs animal food, even a mere diluent is much preferable to wine ; which, whilſt it gives a temporary flow of animal ſpirits, rather oppoſes that neceſſary aſſimilation which nature aims, at in the offices of digeſtion.

Was Coffee ſubſtituted inſtead of the bottle immediately after dinner, it ſeems more than probable that many advantages would flow from it, both to the health of individuals, and general œconomy ; and it ſeems not improbable but by deferring Coffee or Tea ſo late as is uſually practiſed, we interrupt digeſtion, and add a new load of matter to that already in the

5 ſtomach,

ftomach, which, after a full meal, is not a matter of indifference.

On the contrary, ever fince I was capable of forming an opinion on fubjects of this nature, I could not forbear thinking, that the ufe of Tea in an afternoon, at the time and in the manner it has generally been practifed, is exceedingly prejudicial to many perfons ; and if many have efcaped without feeling any prejudicial effects, they may juftly afcribe it to the firmnefs of their conftitution ; I was almoft tempted to fay, to their good fortune. This matter, I own, is capable of much difpute ; and the more fo, as minute diftinctions muft be called to the aid of both parties.

I cannot however conclude thefe remarks, without repeating the fubftance of what I could wifh to inculcate ; that in refpect to real ufe, and as a part of our food, I have no evidence to induce me to think that Coffee is inferior to Tea.

That, in refpect to national œconomy, the benefit of our colonies, and the lives of the feamen, every circumftance concurs to give Coffee the preference. It is raifed by our fellow fubjects, paid for by our manufactures, and the produce ultimately brought to Great Britain.

That the great obftacle to a more general ufe of Coffee is the very high duty and excife

That leffening the duty would not leffen the revenue ; fmuggling would be difcouraged, and an increafed confumption would make up the deficiency to the treafury.

That the planters would be induced to cultivate Coffee with more care, was there a better market for it.

That, as little planters might be enabled to fubfift by raifing Coffee, &c. their numbers would increafe, and add to the

ftrength

strength of the several islands; as Europeans might endure the labour requisite for the cultivation of Coffee.

I have subjoined the translation of a paper communicated to me by Governor Melvill, whose unwearied endeavours to promote the interest of Great Britain and her colonies deserves every grateful acknowledgement; and likewise the copy of a letter I received long since from the late Governor Scott of Dominica; I persuade myself, that evidences like these will have some weight with the public.

Should any part of these remarks afford my friend reason to think they may contribute to the benefit of the community, he is at liberty to make use of them in what manner he pleases.

J. FOTHERGILL.

G   *Observa-*

Observations on Coffee, by a learned and experienced Planter at The Grenades, *communicated to Doctor* Fothergill, *by Governor* Melville.

Tranflated from the French.

SEVERAL perfons in Europe imagine, that a much better kind of Coffee might be gathered in our Iflands, than that which is ufually brought from thence. There is no doubt of this, and our inhabitants are very fenfible of it; but the ever-powerful motive of intereft prevents them from endeavouring to improve the produce of this plant.

They learn from experience, that a light foil, dry and elevated flopes, produce Coffee of a fmaller berry, and more delicate flavour; and that all the Coffee which grows in a low, fertile, and moift foil, is bad, the berry large and flat, and almoft infipid.

Experience alfo teaches them, that trees planted in thefe foils yield commonly from 12 to 16 ounces of Coffee *per* plant; in the other foils they fcarcely furnifh more than from 6 to 8 ounces; this makes an immediate difference of one half in the weight. Now in France, England, and all the European markets, the only ftated difference in the price of the fmall well-prepared Coffee, and that which is larger and of the worft kind, is from 15 to 20 *per Cent.* The inhabitants therefore would neceffarily find it their advantage to plant their Coffee-trees in the richeft foil; and thofe perfons only will have the fmall and fine Coffee, who have no other than bad grounds, and have not a fufficient number of Negroes to manure and improve them.

The

The calculation is eafily made : with an equal number of plants double the weight is produced ; and by the difference of price no more is loft than from 15 to 20 *per Cent*. Intereft hath therefore prevented our inhabitants from applying themfelves to the culture of that kind of Coffee which is moft valued in Europe. To excite a proper emulation among them, the difference of price between the various forts of Coffee fhould be as confiderable as it is between the feveral kinds of Sugar.

To thefe confiderations we may add, that the trees laft a much longer time in the fertile grounds, and that they need not be tranfplanted fo frequently.

Some fkilful perfons have advifed to follow the method of the Arabs, with refpect to the preparation of Coffee, in two particulars ; firft, never to gather it till it is perfectly ripe ; fecondly, to dry it in the fhade, when feparated from the pulp.

The laft of thefe is fcarce poffible ; becaufe, although the air is very-hot in thefe climates, it is always fo damp, that we know from experience the Coffee could never be dried in the fhade fufficiently for exportation into Europe.

The firft would be very ufeful, and even poffible, if other perfons were employed in the bufinefs than Negroes, who, being lazy, ignorant, and generally ill-difpofed, either cannot, or will not, attend properly to this particular ; and have no other wifh but to finifh their work as faft as poffible, either to get rid of the tafk impofed upon them, or to avoid punifhment. Befides, the feafon for gathering the Coffee being near the winter, the rains, which are then very frequent, often make the berries fall before they are perfectly ripe.

As to the hiftory of our Coffee, it certainly comes originally from Babel Mandel. The firft tree that was brought to Martinico in 1727, or 1728, by Mr. Delieu, came from the gardens of his

Moft

Moſt Chriſtian Majeſty, and was of the ſame ſpecies as that which is at Batavia. This tree was planted near to Port Royal, in a fertile moiſt ſoil, and almoſt level with the ſea, ſo that the ſpecies muſt neceſſarily have degenerated.

All the lower claſs of people in Martinico before this time cultivated the Cocoa; but, by a contagion, as difficult to account for, as the effect of it was general, all the Cocoa-trees periſhed in 1727. The inhabitants, half ruined, after having tried ſeveral ſchemes, reſolved at laſt to plant Coffee; and the French Eaſt India Company having lowered the duties, this cultivation was much encouraged.

The French are in general more cautious in the exportation of their Coffee than the Engliſh; they put it into caſks that are very dry: in the Windward Iſlands, where the beſt Coffee is made, the veſſel is neither laden with Raw Sugars, nor with Rum; Clayed Sugars only are exported with it, which are of little detriment to this berry. The captains take care alſo to place it between decks, or in ſome other very dry part of the ſhip. The Engliſh, on the contrary, ſtow Raw Sugars and Rum in almoſt every part of the veſſel. Theſe do a conſiderable injury to the Coffee that lies near them.

There is another more diſtant cauſe, that few people have noticed, but which contributes greatly to the badneſs of the Coffee exported into England.

Moſt of the Engliſh ſhips are hired for the freight; the captains ſtow the goods as they receive them; and the owners are ſatisfied, if the veſſel is but well filled. It is a matter of little concern to them, whether the ſeveral kinds of goods have been properly diſpoſed, or whether they have received any detriment by lying near each other. The French ſhips are generally laden for the proprietors own uſe; the captains buy the

goods

goods themfelves; and, that they may be able to give a proper account of their management, and to fhew that they have acted with prudence and caution, they are obliged to pay great attention to the ftowage of their veffel, and to the prefervation of their cargoes. Hence it follows, that the Coffee which is carried to France is better than that which is brought to England.

Extract

Extract of a Letter from George Scott, *Esq.*
late Lieutenant Governor of Dominica, *to Dr.*
Fothergill.

Government Houfe, Ifland
Dominica, Nov. 21. 1765.

SIR,

MR. I⸺, a confiderable planter of Grenada, touched at this ifland on his way thither; and, in the courfe of his intelligence, having made me acquainted with your patriotic efforts for encouraging the growth and produce of the infant colonies lately ceded to us by France; and being very defirous of throwing in my mite towards forwarding your very laudable labours; I have therefore taken the liberty of putting on board the fhip Neptune, Edmund Stevenfon mafter, under the care of Mr. Beats, of London, in a box directed for yourfelf, three fmall bags of Coffee, which you will do me the favour to accept, trifling as they are, though I have not the honour of your acquaintance, as they are only intended for whatever experiments you may think proper to make of them.

The little bag marked N° 1. was gathered in the year 1760; that marked N° 2. in 1763, and N° 3. laft year: all the growth of this ifland, which is looked upon to make the beft Coffee in the Weft Indies, excepting that of the ifland of Mary Gallant; and on the Weft fide of the ifland of Martinique, on the mountains oppofite the Diamond Rock; which Coffee the French always gave the preference to, though the inhabitants of this ifland

prefer

prefer their own, which they always make ufe of for their breakfaft, taking equal quantities of it and boiled milk (or, more properly fpeaking, milk that is fcalded), and after their dinner they
commonly drink a cup of Coffee without milk; and they have
in general excellent health, and a fine flow of fpirits, for this part
of the world: whereas the Englifh fubjects, whom it is difficult
to wean from prejudices, ftill perfift in the ufe of Tea; and, though
they enjoy a good ftate of health, do not appear to have half the
vivacity or livelinefs with the French in the fame ifland
with us.

I am told, that in England they ftamp a value upon Coffee in
proportion to the fmallnefs and greennefs of the grain; here they
regard neither the fize nor colour of it for their own ufe, but
efteem it in proportion to the time it has been gathered, and for
having been kept in a dry warm place, and expofed to the air
three or four times a year; and the greater number of years it is
kept in this manner, the better it is, they fay. They alfo afcribe
a great deal of its excellence to the method of preparing it for us:
if it is over roafted, it has a flat, bitter, and burnt tafte; and if it is
not roafted enough, though the Coffee fhould be five or fix years
old, it will tafte as if it had been gathered this year: but if it is
old and well roafted, and immediately covered up fmoaking hot
in a bowl or cup, to prevent the fine volatile particles and flavour
from going off: if then, when cold, it is ground and made properly with boiling good water, it is looked upon to be in its
higheft perfection. The better fort of French, in all the iflands,
make a practice of taking a cup of equal parts, Coffee and fcalded
milk, with a cruft of bread, almoft as foon as they get out of bed
in a morning; and the reafons they give for this cuftom are, that
it clears the brain, enlivens the fenfes, cleanfes the ftomach, throws
off any rheum or fortuitous matter that may be lodged about

4

the

the head, ftomach, or lungs, from foul air or putrid vapours. And they likewife fay, that it prevents, and even cures, the gravel. The Turks alfo fet the higheft value upon good Coffee, on account of its exhilarating qualities, and brightening the animal fpirits. Surely then it muft be preferable to Tea, which has quite contrary effects in moft fhapes whatever; and it muft, in my humble opinion, be one of the beft breakfafts in the world, for the honeft, brave people of the foggy ifland of Great Britain, where fuch a multitude of melancholy accidents happen from a lownefs of fpirits. But what effects Coffee or Tea have upon the body or mind, you, Sir, muft be the beft judge, as it is your principal ftudy and profeffion to know the œconomy of the human frame; mine having been ever that of arms, and at prefent to prefide over this ifland; from whence I will with pleafure fend you annually as much Coffee, of whatever fort you like, as you may want for your own ufe, while I remain here; being very much (though unknown)

SIR,

your moft obedient

humble fervant,

GEORGE SCOTT.

P. S. The method of curing Coffee through the Weft Indies is, by paffing it through a mill after it is ripe and gathered; and after this operation, it is put into cifterns and covered with water for ten or twelve hours, until the pulp becomes loofe, when it is wafhed, and the Coffee, being in its hufks, is thrown in heaps to fweat, and that the water may drain off, for two or three days more, when it is fpread abroad and dried in the fun; and when dry, is put into troughs, and pounded with rammers, until all

the

the huſks (or parchment, as they call it) are beat off, when it is winnowed in the air, and expoſed in the ſun until it is perfectly dry, and then carried to market. The Coffee I ſend you, in the little bag marked N° 3, is ſome of a ſmall quantity I got one of the planters to make for me after the method of Mocha; which, as I am informed, is by ſweating and drying it in the ſhade, after it is paſſed through the mill, and muſt, in my opinion, be infinitely preferable to ſoaking it in water, and drying it in the ſun, which certainly muſt extract abundance of its virtues; particularly that fine flavour good Coffee has, which is ſo grateful to the ſmell when it is firſt poured out. This little bag, I believe, you will find very good, though it has not been cured above ten months: and if there was enough of it to keep for three or four years, I imagine it would be perfectly excellent; for this method muſt certainly be the beſt, and I have endeavoured to perſuade many of the planters to come into it; but the great expence they muſt be at to erect buildings to cover it from the ſun and rain, while it would be cur-ing; the great labour and time it takes to cure it after the manner of Arabia, and the ſmall price it bears at preſent, will not as yet permit them to come into it; though I am fully perſuaded that the Coffee of this iſland is full as good, when pulled off the tree, as any in Arabia, was it but cured after the ſame method; which I doubt not the planters will adopt, ſhould the price of Coffee riſe, ſo as to encourage them in the undertaking.

G. S.

H　　　　　　　　　　　*Letter*

Letter from a Merchant of London *to* J. ELLIS, *Efq;*
F. R. S. Agent for Dominica.

Sept. 4, 1773.

DEAR SIR,

I HAVE heard with pleafure, that you are preparing for the
public fome obfervations on Coffee, with a view to promote
in this country a more general confumption of what is produced
of that valuable article in our colonies.

I am perfuaded the Weft India planter will find, in your publication, many ufeful hints for improving the quality of his
Coffee.

But I muft beg leave to remark to you, that it is in vain to
think of extending that trade, while the duties, on and after importation, continue fo very confiderable.

I do not however mean to difcourage you. For though the
times are unfavourable for propofing the reduction of any tax; I
truft there are gentlemen in adminiftration who will countenance
fuch a meafure, when juftice to a part, and the good of the
whole community require it. And I conceive this to be a cafe
of that nature. The duties I have mentioned were voted at a
time when the culture of the Coffee-plant was unknown in our
iflands, and when the confumption was fupplied intirely from
Arabia. It might then be reafonable to confider it merely as an
article of luxury. But circumftances are now greatly altered.
The iflands acquired by the late peace, and Dominica in particular, have large plantations of the Coffee-tree, and the planters are
well fkilled in the cultivation of it. They could furnifh the prefent

fent confumption, and any further quantity that might be wanted. I know that, a few years fince, the excife on foreign Coffee was raifed for the encouragement of the Britifh iflands. But the duty and excife on our own were left as before; which are fo con-fiderable, as to reftrain the middling and common people, who alone make a large confumption, from the ufe of it. The French in this feem to have underftood their intereft better; their Coffee pays but a fmall duty, and Tea is fcarce heard of among them.

It might be fo in this country, did we not make that article, as well as Chocolate, dearer than Tea, by difproportionate and enor-mous duties; which otherwife would be fold as cheap, and pro-bably be the means of preventing, in a great meafure, the expor-tation of our bullion to China [*a*]. This can only be avoided by fubftituting another *focial refrefhing liquor* inftead of Tea: Cof-fee and Chocolate are its natural rivals, and would, in all like-lihood, have the fuperiority, if government would be fatisfied with their contributing to the neceffities of the ftate in the fame proportion. More at prefent is exacted; and that alone difables them from a competition. You will doubtlefs think it ftrange, that articles which our own colonies can raife, fhould pay a higher duty than a Chinefe commodity, the place of which they might fupply. You may perhaps think me miftaken in the affertion. I fhall therefore endeavour to prove it, in fo clear a manner, as may convince not only you, but any man, however little he may be accuftomed to reflection.

[*a*] We fee at prefent, that a temporary fufpenfion of the India Company's purchafes of Tea has confiderably affected the price of filver; fo that we may foon expect a new coinage, which could not have happened, had they continued to drain this kingdom of bullion as formerly.

For

For that purpose I muſt ſtate the following plain matters of fact:

One eighth part of an ounce of Tea, that is, one ſpoonful and a half, is commonly uſed for the breakfaſt of one perſon. At that rate, a quarter of a pound is conſumed in 32 days; which, to avoid fractions, I ſhall conſider as a month, both with reſpect to the other articles and this: ſo that upon the whole it will make no difference. A quarter of a pound per month, is three pounds in the year.

One quarter of an ounce of Coffee is uſually allowed for a good diſh; and I might very well ſuppoſe that, were it cheap, three ſuch diſhes would be conſumed for a breakfaſt. However, to avoid objections, I ſhall reckon but two; which will require half an ounce of Coffee; that is, *four times* the weight of the Tea; conſequently one pound in a month, and twelve pounds in the year.

I am informed, that it is common to give out one of the ſmall diviſions in a cake of Chocolate, of which there are eight in a quarter of a pound, to make one diſh: two, at leaſt, would be requiſite for a breakfaſt; and they would weigh an ounce; which is *eight times* as much as the Tea, and *double* the weight of the Coffee. The conſumption of the month would be two pounds; and of the year, twenty-four.

From hence it is plain that, if Tea is charged with duties and exciſe to the amount of 2s. 10½d. per pound, which is actually the caſe, as I ſhall ſhew preſently, roaſted Coffee, of which *four times* the quantity is neceſſary for the ſame purpoſe, ſhould pay but one *fourth part* of that ſum; that is 8 d. and ⅝ths per pound; and Chocolate, one *eighth part* being 4 d. and $\frac{5}{16}$ths; and if the duty and exciſe ſhould continue to be paid on the Coffee before it is roaſted, they ought to be near one quarter leſs than I have men-tioned,

3

tioned, becaufe it lofes of its weight in roafting 24 lb. on 112 lb. The lofs of weight on the Chocolate nut is likewife 18 lb. on an hundred.

Allowing for which, the duty on roafted Coffee will be reduced to 6 ¼ d. and Chocolate fhould not pay quite 3 ¾ d. inftead of 33 s. 6 d. per hundred on plantation Coffee at the cuftom houfe; that is 4 d. per pound, and 1 s. 6d. per pound at the excife; in all 22 d. It muft be afterwards roafted; which reduces 112 lb. to 88 lb. and 22 d. upon the former, is full 2 s. 5 d. on the latter.

On landing Chocolate nuts, 11 s. 11 ½ d. per hundred is paid; which is 1 ¼ d. per pound; and the excife on the Chocolate, when made into cakes, is 2s. 3d. per pound more. Therefore the duty upon a pound of this article, is nearly the fame as on Coffee; though *double* the quantity is required for a breakfaft.

The duties on Tea are as follows, 25 *per cent. ad valorem,* paid by the Eaft India Company, and as much by the buyer; making together 50 per cent.; and that, on the average value of Tea, is 22 ½ d. per pound: for, according to the beft information I have been able to procure from the Tea-brokers, 3 s. 9 d. is the medium price at the fales; the high-priced forts, the hyfon and fouchon, not being a tenth part of the importation. Befides the above duties, there is 1 s. per pound excife; in all, 2 s. 10 ½ d. per per pound on 3 s. 9d. value; which is *eighty per cent.* While plantation Coffee, which is rated at 15 d, though in reality it fells but for 6 d; and Chocolate nuts, that are nearly of the fame value, pay 2 s. 5 d. per pound, which is *four hundred and eighty per cent.*

I think nothing more is wanting to prove my affertion with refpect to the duties, but the bringing into one point of view, the fums that government would receive on each perfon's annual confumption, upon the footing I have propofed.

On

On 3 lb. of Tea.

The confumption of a year.

	£.	s.	d.	£.	s.	d.
Valued 3 s. 9 d. per pound, which is	0	11	3			
The duty and excife at 2 s. 10½ d. amount to				0	8	7½

On 15 ¼ lb. of unroafted Coffee; which, when fit for ufe, would be reduced to 12 lb.

	£.	s.	d.	£.	s.	d.
The prime coft at 6 d. per pound, is	0	7	7½			
Suppofed duty on the 15¼ lb. at 6¾ d. per pound is				0	8	$6\frac{15}{15}$

On 29 ¼ lb. of Chocolate nuts, called, in the Book of Rates, Cocoa nuts; which would make 24 lb. of Chocolate.

	£.	s.	d.	£.	s.	d.
The prime coft of the 29 ¼ lb. at 6d. per pound, is ——— ——— ——— }				0	14	7½
Suppofed duty thereon at 3½ d. per pound ———				0	8	6

The duties payable at prefent on the fame quantities of the two laft articles ftand thus:

	£.	s.	d.
On 15 ¼ lb. of unroafted Coffee, equal to 12 lb. when fit for ufe, at 1 l. 13 s. $6\frac{3}{20}$ d. per cent. which is 4 d. per pound. ——— ——— }	0	5	1
Excife on the fame, at 1 s. 6 d. per pound. ———	1	2	10½
	1	7	11½

On

	£.	s.	d.
On 29¼ lb. of Cocoa nuts, at 11 s. 11½ d. per hundred; that is, 1¼ d. per pound. ———	0	3	½
Excife on that quantity made into Chocolate, producing 24 lb. at 2 s. 3 d. per pound. ——— }	2	14	0
	2	17	½

I believe the quality of Weft India Coffee might be greatly improved, if the planter were encouraged by a confiderable demand from hence. One great reafon of its being inferior to Mocha at prefent is, that the Dutch and Germans, who are almoft the only buyers, have an unaccountable regard to the colour: they prefer a light green; and to have that quality, the Coffee muft be frefh. The planter, having fcarce any other market, is obliged to conform to their tafte, and haftens to fend away his produce; which may poffibly heat in the paffage, and certainly cannot have had time to get rid of the rank flavour natural at firft to it. The prices which they give are, befides, fo low, that Coffee is become an article not worthy the attention of a planter; and unlefs adminiftration grants it fome confiderable encouragement, I have no doubt but the trees will be rooted up throughout the iflands, and the Coffee trade loft to this nation.

Two years ago, the Germans readily gave double the price for it: but various caufes have concurred to lower the value: fome are of temporary nature; fuch as the high price of corn all over Europe, and the troubles of Poland: others are of a worfe kind, and muft greatly affect the confumption.

The

The king of Pruffia, has, I am affured, laid an additional duty on it of 50 *per cent* ; and the Landgrave of Heffe has, in a great meafure, prohibited the ufe of it.

Another circumftance to difhearten our planters is, that the Dutch of Surinam, and the Berbices, have amazingly encreafed the number of their Coffee trees; and with the advantages of a rich foil, and lands cheaper; an eafy communication by water, and more facility in borrowing money for fuch extenfive under-takings, are better enabled to continue the trade. The French too import aftonifhing quantities from the iflands of Bourbon and Mauritius; where flaves can be had cheap from Madagafcar.

Notwithftanding all this, the feafonable interpofition of govern-ment may yet not only fave the Coffee plantations in our iflands from ruin, but put them upon a more flourifhing footing than ever. No-thing more isneceffary, than to regulate the duties upon an equita-ble footing; and if Coffee and Chocolate fhould pay a proportionable duty with Tea, I think I may venture to affert that the revenue cannot be leffened by the former taking place of the latter. On the contrary, as the Weft India produce is all paid for in our ma-nufactures, the ingenious artift, the tradefman, and hufbandman at home, muft be better enabled to pay taxes. What employ-ment the Weft India trade furnifhes to our people, will appear from the lift I fend you of the articles exported; in which every one will find himfelf interefted directly, or indirectly.

There are other confiderable advantages accruing from the con-nection of North America, Ireland, and Africa, with the iflands, of ftill more confequence to the parent ftate. But to trace them with any degree of precifion, would require more time than I can fpare at prefent, and take up too much of yours.

Certain it is, that the profits of our Weft India fettlements, through different channels, center in Great Britain; where the

planter

planter retires, when his fortune is made, to recruit an exhausted constitution, and enjoy his gains dearly earned by continual dangers and anxiety.

Our unhappy adventurers in Coffee, in the Ceded Islands, begin as I am told, to lose all hope of that reward for their labours, which used to support them under every disappointment, a prosperous return to their family and friends. Their credit is totally stopt by the difficulties of the times, and their produce yields them only one half of what it did in 1770: by which the value of their estates is lowered in the eyes of their correspondents, who, if able, are afraid to assist them, thinking their situation desperate. Their losses in Negroes, and mules, have been immense from the difficulties attending the cultivation of islands overgrown with woods, consequently damp and unhealthy: from the want of provisions, and of proper shelter for their Negroes and cattle. In short, their affairs are at such a crisis, that, unless they have immediate relief, from the wisdom and justice of Parliament; it is scarce possible but they must sink under their misfortunes, and soon be destroyed by the harpy claws of lawyers and usurers.

On the other hand, should the planter be supported and encouraged, in proportion as his produce may encrease; the revenue of the islands, arising from the 4 ½ *per cent.* on that produce, must become so much the more considerable. The planters, who have purchased estates of government, payable in instalments, may then be able to fulfil their engagements; and the lands surrendered by the Caribees of St. Vincent find purchasers at a good price; which cannot be expected in the present situation of things. It has been very well observed in the Dominica petition to paliament, delivered in by you the last session, that Coffee and Cacao plantations deserve particular encouragement; because they can be undertaken upon a smaller scale than sugar estates, which must necessarily be extensive and require expensive buildings. Whatever tends to divide the

I

landed

landed property, and consequently furnish subsistence to a greater number of white inhabitants, certainly adds to the strength of the islands, and enables them the better to resist a foreign force, or to quell insurrections of the Negroes at home.

I am aware it may be objected, that though I have shewn that if the duties and excise on Coffee were reduced to 6 ¾ *d.* *per* pound, and on the Chocolate nut to 3 ½ *d.* and those articles were to supply, in some measure, the place of Tea; the loss of the revenue on the latter would be compensated by an increase of revenue on the former; yet the difference between those duties I have proposed, and what are now actually paid, would be so much lost to the government, on the present consumption of Coffee and Chocolate.

I own that if *all* that is consumed paid the excise and duty, such a loss must be expected: but there is no doubt, but immense quantities of Coffee are smuggled all round this kingdom, particularly on the Southern and Western coasts, where I am assured it may be had, even in small quantities, at the rate of 14 to 18 *d. per* pound, which is little more than half the duty and excise. This is the more detrimental to England, as such Coffee is chiefly the produce of the French islands, and paid for in ready money.

It is very probable, that three times as much West India Coffee finds its way in that manner, as what pays the duties and excise. If so, a reduction of three parts in four of those duties would occasion no loss to government, provided the whole of the consumption were regularly entered; and in all likelihood that would happen, as there would be then no temptation left for smuggling.

But should I grant that, on lowering the duties so that every one might be induced to pay them, there would still be a deficiency: surely considerations of policy and justice to the colonies may reasonably be supposed to have more weight with a

British

Britiſh Parliament, and Adminiſtration, than a difference of no
conſequence upon one of the branches of the revenue.   However,
to prevent all objections, I ſubmit the following plan, for com-
penſating to Government any difference which may be ſuppoſed
to ariſe from leſſening the duties in the manner I have mentioned,
and likewiſe to put an entire ſtop to the ſmuggling of Coffee and
Chocolate.   And firſt I muſt deſire it may be obſerved, that a
great part of the preſent conſumption of thoſe articles is in
Coffee-houſes.

My propoſal is, to let the Cuſtom-houſe duties on plantation
Coffee remain as they are.

To convert the exciſe now paid upon foreign Coffee into a
duty, not to be drawn back upon exportation : for that would
certainly give occaſion to great frauds.   This will make no difference to Government on Mocha Coffee.

To take off the whole exciſe on Coffee and Chocolate, and
make up any ſuppoſed deficiency on the Plantation-Coffee and
Chocolate by a proportionable tax on licences for keeping a Coffee-houſe, or any houſes where Coffee and Chocolate may be ſold
ready made.   Perſons who keep ſuch houſes will have no reaſon
to complain of the tax, as they will be able to purchaſe theſe articles, of Britiſh growth, at a very low rate ; nor will it be any
hardſhip upon their cuſtomers, becauſe there will be no occaſion to
raiſe the price to them.

Another conſideration in favour of this plan is, that the pay-
ment of the tax cannot be evaded, and will be eaſily levied.

I have no doubt, that when the attention of ſo experienced a Finan-
cier as the preſent Miniſter is directed to this object, our fellow-
ſubjects in the Weſt Indies will be enabled to continue the cultivation of the Coffee and Cacao, or Chocolate nut, with as great
advantage to their mother country as themſelves.   But I have
already exceeded the bounds of a letter ; and ſhall therefore conclude with aſſuring you, that, if what I have written, can be of

an

any service to you in your views of promoting the interest of the island for which you are Agent, and of the community at large, I shall not think my time has been ill employed; and that I am, with the sincerest regard,

DEAR SIR,

Your most obedient humble Servant.

POSTSCRIPT.

After having written the above, it has occurred to me, that it might be proper to consider what difference the increased consumption of Coffee and Chocolate, in the lieu of Tea, would make with respect to navigation. The East India Company has lately agreed to pay 26*l.* 10*s. per* ton for freight from China. One ton answers the same purposes as four tons of Coffee; and the freight on that quantity, at the usual price of 5*l.* 10*s. per* ton, would be 22*l.* which is something less than on the proportional quantity of Tea. But the difference is very considerably in favour of Chocolate; and there are most material advantages attending the West India navigation, in preference to the East; particularly, that our brave sailors are less liable to fatal distempers in such a voyage, having no occasion to remain for so long a continuance aboard a ship; and, what is of great importance, should their King and Country, on a sudden emergency, need their assistance, it may soon be commanded. On the contrary, when once a ship for China has left our ports, we can expect no service of that kind from the crew for eighteen months, however they may be wanted.

Exports

Exports to the British West India Islands ; *in which some of the minutest Articles are inserted, to shew that the Planter has Recourse to this Country, even for them.*

TIMBER for Dwelling Houses, Sugar Houses, Distilling Houses, Curing Houses, and Ware Houses, ready squared and framed, to put up on arrival; with Window Frames and Sashes, glazed. But the chief Orders for these are sent to North America.

Bricks.

Tiles.

Lime.

Flat Paving Stones.

Mill Stones.

Mill Cases.

Gudgeons, Carpooses, and Mill Wedges.

Stills for making Rum.

Vatts to contain the Liquor.

Casks or Puncheons to put it in.

Iron Teaches for boiling the Cane Juice.

Wine Sieves to strain it through.

Lime, to refine it.

Skimmers and Ladles.

Pumps for the Still House.

Iron Work for the Fire place under the Coppers, or Teaches.

Coals for the same.

Sheet Lead to line the Cisterns.

Lead Pipes.

Solder.

Solder.

Sheets of Copper, for drying Magnoe, or Caſſava Bread.

Heſſens, for Cotton Bags.

Brown Roll, for Coffee Bags.

Sail Cloth, for ſmall Veſſels.

Fiſhing Tackle.

Ropes and Cordage.

Coaches.

Chaiſes.

Harneſſes.

Fire Engines.

Leather Buckets.

Trunks.

Collars and Traces.

Weights and Scales.

Truſs Hoops.

Carts.

Waggons.

Wheel-barrows.

Horſes.

Mules.

Shoes for them.

Oats.

Beans and Bran.

Saddles.

Bridles.

Surcingles.

Whips.

Spurs.

Boots.

Shoes.

Buckles.

Shoe

Shoe Buckles.

Blacking Balls.

Stockings, of Silk, Thread, and Cotton.

Stocking Breeches.

Fuſtian,

Jean,

Drabs,

Damaſcus,   } for Breeches and

Nankeen,   Waiſtcoats.

Ticks,

Dimities, and a variety of ſtriped Goods,

Thin Woollen Cloth,

Kerſimeers,   } for Coats.

Duroys, and

Thickſet,

English, Iriſh, and Scotch Linen, for { Shirts, Sheets, Table Cloths, Napkins and Towels.

Cambrick, Muſlin, and Lace, for Ruffles.

An immenſe Quantity of Printed Cottons, and Callicoes, and Linen, for Gowns, Night-Gowns, and Petticoats for Houſe Negroes, and White People; Handkerchiefs, of which a great deal for Negroe Women's heads.

Oſnaburgs, and

Checks, for cloathing the Field Negroes,

Jackets for the Men,   } of coarſe Woollen, } In great quantities.

Petticoats for the Women,

Bed Ticks.

Bedſteads.

                                           Hair

Hair Mattraſſes.
Scotch Gauze for Curtains.
Chairs.
Tables.
Commodes, and Cabinet Ware of all kinds.
Hats, for White People and Houſe Servants.
Negroe Hats.
Negroe Caps.
Paper, for hanging Rooms.
Linſeed Oil, and Painters Colours.
Mops.
Brooms.
Scrubbing Bruſhes.
Chalk, and Whiting.
Candle Boxes.
Pepper Boxes.
Common Tinder Boxes.
Piſtol Tinder Boxes.
Kettles.
Pots.
Pans.
Dairy Pans.
Plates.
Diſhes.
Trenchers.
Hand Baſons.
Tea Pots.
Cups and Saucers.
Baſons.
Horn Tumblers.
Silver Rummers, and Goblets.
Glaſs ware,
Table Knives.

2

Forks.

Forks.
Spoons.
Lamps.
Candlesticks.
Lantherns.
Hand Coffee Mills.
Coffee Pots.
Chocolate Pots.
Tea Canisters.
Hour Glasses.
Watches and Watch Chains.
Clocks.
Plantation Bells.
Hand Bells.
Swords and Hangers.
Pistols.
Guns.
Gun Powder, to split Rocks.
Shot.
Brass Cocks.
Bolts.
Hinges.
Locks.
Padlocks.
Nails of all Sizes.
Crows.
Hammers.
Hatchets.
Axes.
Saws.
Gimblets.

K

Files.

Files.
Chifels.
Planes.
Rules and Lines.
Hoes and Bills in great quantities.
Shovels.
Spades.
Rakes.
Garden Sheers.
Watering Pots.
Grind-ftones, bored and with Handles.
Combs.
Brufhes.
Hair Powder.
Powder Puffs.
Razors.
Shaving Boxes, with Soap and Brufhes.
Hones and Straps.
Needles.
Thread.
Sewing Silk.
Tape.
Laces.
Hooks and Eyes.
Cap Wire, and Tags.
Pins.
Thimbles.
Ribbands.
Garters.
Clafp Knives.
Buttons.
Gold and Silver Lace.
Pen Knives.

Pens.

Pens.
Ink Powder.
Ink Horns.
Standiſhes.
Paper.
Books for Accounts.
Wax.
Wafers.
Seals.
Almanacks.
Books.
News Papers.
Magazines.
Playing Cards.
Meſſage Cards.
Dice.
Back Gammon Tables.
Chirurgical Inſtruments.
Hydrometers.
Proof Bottles.
Pipes.
Tobacco, prepared for ſmoaking.
Snuffers.
Salvers.
Bottle Stands.
Bottle Tickets.
Snuff.
Snuff Boxes.
Garden Seeds.
Manna.
Salts.
Jalap.

K 2

Bark.

Bark.
Ippecacuanha.
Salfepareille.
Nitre.
Camphire.
China Root.
Rhubarb.
Tartar Emetic.
Cream Tartar.
Vitriol.
Verdigrife.
Antimony.
James's Powders, and Medicines of all kinds.
Eau de Luce, and fweet-fcented Waters.
Pepper.
Cloves.
Mace.
Durham Muftard.
Ketchup.
Pickled Cucumbers.
Pickled Walnuts.
Vinegar.
Salad Oil.
Olives.
Almonds.
Raifins.
Currants.
Figs, and other dried or preferved Fruit.
Sago.
Pearl Barley.
Split Peas.
Grutts.

Lamp

Lamp Oil.
Refined Sugar.
Tea.
Soap.
Wine.
Porter in Casks.
Porter in Bottles.
Gloucester Cheese.
Cheshire Cheese.
Hams.
Bacon.
Smoaked Beef.
Tongues.
Butter.
Salt Beef. } These three chiefly from Ireland.
Pork.

They receive besides, every year, a considerable quantity of Herrings from Scotland; and abundance of Flour, Rice, Live Stock, Cod, and Lumber, from North America; and Negroes to an amazing amount from Africa; which are all brought in British vessels, and chiefly purchased with British manufacturers.

TABLE

TABLE

OF THE

CONTENTS.

FINIS.

The Cask for sowing East India seeds with
the openings defended by Wire.

The Box with West India and W. Florida plants shut down
with the openings at the ends and front left for fresh Air.

The Box with divisions for sowing different
seeds in earth & cut moss from the southern Colonies and the West Indies.

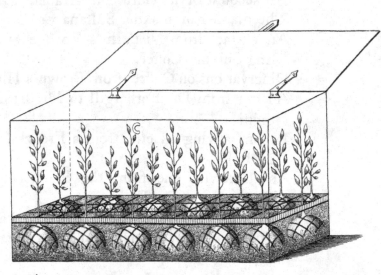

The Inside of the box shewing the manner of securing the roots of W. Florida
and W. India plants surrounded with earth & moss tied with packthread
and fastend crofs & crofs with laths or packthread to keep them steady.

DIRECTIONS

FOR BRINGING OVER

SEEDS AND PLANTS,

FROM

THE EAST INDIES

AND

OTHER DISTANT COUNTRIES,

IN

A STATE OF VEGETATION:

TOGETHER WITH

A CATALOGUE of such FOREIGN PLANTS as are worthy of being encouraged in our AMERICAN Colonies, for the Purposes of MEDICINE, AGRICULTURE, and COMMERCE.

TO WHICH IS ADDED,

The Figure and Botanical Description of a new SENSITIVE PLANT, called

DIONÆA MUSCIPULA:

OR,

VENUS's FLY-TRAP.

BY

JOHN ELLIS, F. R. S.

LONDON,

Printed; and sold by L. DAVIS, Printer to the Royal Society, opposite Gray's-Inn, Holborn.

MDCCLXX.

Directions for Captains of Ships, Sea Surgeons, and other curious Persons, who collect Seeds and Plants in distant Countries, in what Manner to preserve them fit for Vegetation.

IT might be reasonably supposed, from the great quantity and variety of seeds which we yearly receive from China, that we should soon be in possession of the most valuable plants of that vast empire; yet it is certain, that scarce one in fifty ever comes to any thing, except a few varieties of annual plants, which have been common in our gardens for many years. The intention of those who purchase or collect these seeds is, without doubt, to oblige the curious in these kingdoms, by procuring what they suppose may prove both ornamental and useful: but how contrary to their intentions do their friends find it, who, being under great obligations for this expensive present, have the mortification to be totally disappointed in their expectations! These remarks are therefore intended to prevent, if possible, the like disappointments for the future.

The crafty Chinese traders, perceiving that many of the Europeans who buy these seeds are very little acquainted with the nature

B   of of them, take the advantage of their want of knowledge; and, in order the better to deceive them, put up a great variety of forts in a very neat manner: when the feeds arrive here, and come to be examined by perfons of judgement, they foon find that moft of them have been collected many years; confequently are decayed, and of no value. To prevent this fraud for the future, it would be proper to examine the ftate they are in before they are purchafed. And though it is very difficult to judge how long they may have been gathered, yet we may form a tolerable judgement of them by cutting fome of the larger ones acrofs, and bruifing the fmaller ones: By the help then of a magnifying glafs of two inches focus, we may difcover, whether their internal part, which contains the feminal leaves, appears plump, white, and moift. If fo, thefe are good figns of their being in a vegetating ftate; but if they are fhriveled, inclining to brown or black, and are rancid, they cannot in the leaft be depended upon.

The refident factors in China are the propereft perfons to collect the choiceft kinds; they will follow any ufeful hints with chearfulnefs. Many valuable trees, unknown in Europe, grow in the northern provinces of China; the feeds of thefe may be obtained by means of the miffionaries at Pekin: that climate, though in 40 degrees of North latitude, is liable to more fevere cold than ours in winter. So that trees from thence would thrive well with us in the open air, but much better in the fame latitude of North-America, on account of the great heat of the American fummers. The Secretary of the Royal Society of London correfponds with the Miffionaries; and there is no doubt but, upon a proper application, they would with pleafure oblige the Society, as they have done formerly, in fending many curious feeds. But as the diftance is great, and the manner of preferving the feeds properly, fo as to keep them in a ftate of vegetation, is an affair of confequence, the

following

following hints may be of use in bringing them over to answer the end proposed.

In the first place it ought to be carefully attended to, that the seeds should be perfectly ripe when they are gathered; and they should be gathered, if possible, in dry weather; afterwards they should be spread thin on paper or matts, in a dry airy room, but not in sunshine. The time necessary for this operation will vary according to the heat of the climate, or season of the year, from a fortnight to a month, or perhaps two may be necessary; the hotter the season, the less time will suffice. This is to carry off their superfluous moisture, which if confined would immediately turn to mouldiness, and end in rottenness.

As there are two methods that have succeeded, and put us in possession of several young plants of the true tea-tree of China, I shall mention them both, in order to assist the collector in bringing home the seeds of many valuable plants.

The first is by covering them with bees-wax in the manner explained in the Phil. Transact. vol. LVIII. p. 75. and which is hereafter described; where the acorns vegetated freely after they had been kept a whole season inclosed in wax*.

* Here we must observe, that, in the experiment made on the oak acorns inclosed in wax, they were not put into it till the latter-end of February, though they had been ripe and fallen from the tree four months before, which was the latter-end of October preceding; not but that they might have been safely inclosed much sooner.

However, by this time, that property, which all living substances, as well animal as vegetable, of imbibing and perspiring, was very much abated; for the seeds of vegetables, like animals in their torpid state, do imbibe and perspire to a certain degree; yet this degree greatly diminishes in proportion to the time they are kept (under certain circumstances of the manner in which they are kept) till at last they lose their vegetating power. So that we see how necessary it is, that the larger seeds, that are intended to be inclosed in wax, should be in such a state, as not to send forth too great a quantity of aqueous moisture, and yet that there should be sufficient to support them in this confined state. Many of the tea-seeds lately sent over in wax have perished for want of this caution.

Skilful persons, by cutting some of them open and observing the state of the kernels, will be able, after different experiments, to hit on the critical time for this operation.

It

It principally confifts in choofing only fuch feeds as are perfectly found and ripe. To prove this, we muft cut open fome of
them to judge what fituation the reft may be in, taking care
to lay afide any that are outwardly defective, or marked with
the wounds of infects. When a proper choice of them is
made, they fhould be wiped extremely clean, to prevent any dirt
or moifture being inclofed; each feed then fhould be rolled up
carefully in a coat of foft bees-wax half an inch thick: the deep
yellow Englifh bees-wax is the beft. When you have covered
the number you intend to inclofe, pour fome of this bees-wax
melted into a chip-box of fix or feven inches long, four broad,
and three deep, till it is above half full; and juft before it begins
to harden, while it is yet fluid, put in the feeds you have rolled
up in rows till the box is near full; then pour over them fome
more wax while it is juft fluid, taking care when it is cold to ftop
all the cracks or chinks that may have proceeded from the fhrink-
ing of the wax, with fome very foft wax; then put on the cover
of the box, and keep it in as cool and airy a place as you can.

The method of inclofing tea-feeds fingly in wax, and bringing
them over in that ftate, has beeen practifed for fome time; but few
have fucceeded, owing to the thinnefs of the coat of wax, or
putting paper firft round them, or inclofing them too moift.

The ftones of mangoes have been covered in the fame manner,
but moft of them have been pierced by infects through the wax,
and of many of them that were not pierced, their kernels were
black and hard; a plain proof they had been too dry before they
were inclofed, and that thefe large ftones require as ftrong a
covering of wax as the oak acorns, to prevent the air or infects
coming to them.

It appears from experiments made by curious perfons in the
Eaft-Indies, that mangoes will vegetate fooner by fowing only
the

the kernels: if then some of the ripest kernels were taken out of the stones cautiously, without bruising them, and preserved in the same manner as the oak acorns, it would be an experiment worth trying, in order to obtain this most valuable tree, especially if some of these kernels so preserved were taken out of the wax at St. Helena, and sown in boxes of fresh earth. The same might be practised with success on the tea-seeds, as some of my friends, who have taken this hint, have experienced very lately.

The second method that has been tried with success is, by procuring the tea-seeds in their pods or capsules, when they are brought down fresh from the tea-country at the latter-end of the year, to Canton, at the time that our East-India ships are preparing to depart for Europe. The seeds then in their pods are to be put into pound or half-pound canisters made of tin and tutenague *, with a double rim to the top: the inside of the canister should be first lined with silk paper, or the paper commonly used in China, and the seeds pressed down close, but not so as to be bruised. When the canister is near full to the neck, some more of the same paper must be stuffed in very close, till it is full to the top, and then the double-rimmed cover should be put on very tight. Care must be taken that the seeds are not too moist when they are put into the canister, and that they are found and in good order. The canister then is to be kept in an airy cool place. If the ship arrives early in England, I mean in June or the beginning of July, they may be sown with success; the sooner it is done, the better chance we shall have of

---

* Whether there is any particular antiseptic quality or power of resisting putrefaction in the metallic parts of these kind of canisters, I will not pretend to determine; but it is most certain, that there are sulphureous mineral steams, very perceivable to persons of a nice sense of smelling, that are continually flowing from almost all metallic substances, especially in hot weather; which steams being confined, may probably resist putrefaction, and destroy insects in vegetable bodies; and perhaps these may rather promote than impair their vegetative powers, as I shall shew hereafter in an instance of the use of common sulphur applied for this purpose; for most of the tea-seeds had pushed forth roots in the canister.

their

their growing. Those seeds which I have seen brought home in this manner, had shot out roots, owing to the heat of the climates they had passed through, and the confined moisture; and though not above twenty out of two hundred in the canister succeeded, yet these are thought a great acquisition. Perhaps there would be less danger of so many of them putrefying, if each capsule with its seeds was wrapped up tight in a separate piece of paper, and afterwards closely packed in the canister as before-mentioned. We see how long oranges, lemons, and other fruit wrapped up singly in papers, and put into close packages, will continue sound by the papers absorbing the moisture that must exsude from them, and which prevents their heating and putrefying.

Tea-seeds, put up in this manner, require less trouble than those that are rolled up singly, and afterwards immersed in melted wax. Experience will determine which is the best method. When the ship arrives at St. Helena, they may be easily examined to see in what state they are, by cutting some of them open ; and if they are found, some of them should be sown immediately in cases or tubs of fresh earth, well secured from rats, and the vacancy made in the canister immediately filled up, and stuffed very close with the same sort of paper, to prevent the air getting to the rest, which would soon spoil them. These that are thus put into earth should have but little water given to them till they pass the tropic of Cancer; otherwise they will spire up very weak, from the great heat, and want of a free circulation of good air.

It might be proper, after the ship has passed the tropic of Cancer, near the latitude of 30 degrees North, to sow some more seeds in the same boxes, by which experiment we may judge the better of the properest place to sow the seeds at sea. It has been practised by many, to sow the seeds immediately on leaving China; but this is rarely attended with success, particularly on account of the bad weather too often met with in coming round the Cape of

Good

Good Hope; befide, the young plants are apt to grow too freely and flender in their confinement, and therefore lefs able to bear the cold air when they arrive in this latitude.

If by chance the tops of fuch plants as come up fhould be broken off by any accident, the earth and feeds fhould not be thrown away, for the remaining part of the ftem next to the feed will fhoot up afrefh, as I have experienced in the pot of oak acorns (that grew after they had been preferved a feafon in wax); fome of the tender young plants were by accident broke off fhort clofe to the earth; but before the fummer was over they grew up again, full as vigoroufly as thofe that were not hurt.

The following is a defcription of a proper-fized box to fow the feeds in, in the Eaft-Indies or on the voyage. It fhould be three feet long, fifteen inches wide, and eighteen or twenty inches deep, or more, as it may be found convenient, with a proper cover of wire to fecure the feeds or young plants from vermin, and a lid with hinges to fhut down over the wire, as there may be occafion, and a handle at each end, to move the box eafier to and fro. The ends of the box near the top muft be bored full of holes, to let the crude vapours pafs off that arife while the cover is obliged to be let down; or a fmall valve or wooden fhutter at each end to open outwards, of fix inches long and three broad; the openings to be defended with wire, to prevent the rats getting into the box. This hint is fufficient to fhew that air is abfolutely neceffary, and an ingenious carpenter will eafily contrive fmall doors or openings all round for the health of the young plants.

Or a cafk, perhaps, may be made equally as convenient for this purpofe, as the cooper on board a fhip has always fpare cafks more ready than boxes. The following is the proportion it fhould be of: two feet three inches high, two feet bung diameter, and one foot nine inches head diameter; there fhould be a large opening at the top wired over, the wired part of which might be lifted up at pleafure, and a lid with hinges to cover it; this may

2 be

be either circular or square, as will be moſt convenient, the larger the better; and on the upper part of the ſides there may be four or five little openings wired, with doors to each, for the ſake of giving air all round upon ſome occaſions. Care muſt be taken not to expoſe the young plants to ſtrong ſunſhine: ſometimes, when the lid and doors are open, it may be neceſſary to throw a matt or thin cloth over them, but this muſt depend on the judgement of the perſon who has the care of them; there ſhould be handles fixed to the ſides, to move it with more ſafety.

There ſhould be a layer of wet moſs, of two or three inches deep, at the bottom of the box or caſk; or, if that cannot be got, ſome very rotten wood or decayed leaves, and then freſh loamy earth, about twelve inches deep, both of which will ſink to a foot deep: the wet moſs is intended to retain moiſture, and to keep the earth from drying too ſoon.

The ſurface of the earth ſhould be covered with moſs cut ſmall, which now and then on the voyage ſhould be waſhed in freſh water, and laid on the earth again to keep the ſurface moiſt, and to waſh off mouldineſs or ſaline vapours which may have ſettled on it. When the plants come up, it will be proper to ſave what rain-water can be got, which will encourage their growth, and be of more ſervice than the water drawn out of caſks that have been long on board the ſhip.

Theſe kind of boxes or caſks will be very proper to ſow many ſorts of ſuch ſeeds in as are ſo difficult to be brought from China, and other parts of the Eaſt-Indies, to Europe in a vegetating ſtate; ſuch as the lechee, mangoes, mangoſteens, pepper, marking nuts, various ſorts of peaches, roſes, oranges, citrons, lemons, &c.

And nothing can be more convenient than theſe caſks, for ſending olive plants, capers, young vines, &c. &c. to our diſtant American plantations. The ſize may be varied as the plants to be ſent may require.

To

To this I muft add a method that promifes fuccefs for bringing over plants from the Weft-Indies, and the fouthern parts of North-America, particularly Weft-Florida, the voyage from hence being longer than from the Weft-Indies, and more attention is required to keep the plants in health, than from any other parts of our North-American fettlements: but as there is a good deal of difference in the climates of thefe places, it will be neceffary to obferve, that plants from the Weft-Indies fhould be put on board in the latter-end of Spring, fo as to arrive here in warm weather, otherwife they will be deftroyed by the cold of this lati-tude; and the ever-greens, which are the moft curious from Weft-Florida, muft be fent in the winter months, while their juices are inactive, fo as to arrive here before the heats come on. If the plants fent from thefe countries were planted in pots or boxes, and kept there a year, they might be brought over with very little hazard; or even if they were firft tranfplanted from the woods into a garden, till they had formed roots, they might be fent with much more fafety.

The fize of the boxes that will be moft convenient for ftowing them on board merchant-fhips, where there is very little room to fpare, fhould be three feet long, fifteen inches broad, and from eighteen inches to two feet deep, according to the fize of the young trees; but the fmalleft will be moft likely to fucceed, pro-vided they are well rooted. There muft be a narrow ledge nailed all round the infide of the box, within fix inches of the bottom, to faften laths or packthread to form a kind of lattice-work, by which the plants may be the better fecured in their places. If the plants are packed up juft before the fhip fails, it will be fo much the better.

When they are dug up, care muft be taken to preferve as much earth as can be about their roots; and if it fhould fall off, it muft

C be

be fupplied with more earth, fo as to form a ball about the roots of each plant, which muft be furrounded with wet mofs, and carefully tied about with packthread, to keep the earth about the roots moift: perhaps it may be neceffary to inclofe the mofs with fome paper or broad leaves (as the palmetto) that the packthread may bind the mofs the clofer. Loamy earth will continue moift the longeft. There muft be three inches deep of wet mofs put into the bottom of the box, and the young trees placed in rows up-right clofe to each other, ftuffing wet mofs in the vacancies between them, and on the furface; over this palmetto leaves, if to be had, fhould be put to keep in the moifture, and over them the laths are to be faftened crofs and crofs to the ledges or pack-thread to be laced to and fro, to keep the whole fteady and tight. The lid of the box fhould be either nailed down clofe, or may have hinges and a padlock to fecure it from being opened, as may be found neceffary, with proper directions marked on it to keep the lid uppermoft. There muft be two handles fixed, one at each end, by which means there will be lefs danger of difturb-ing the plants. Near the upper part of the ends of the box, there muft be feveral holes bored to give air: or in making the box there may be a narrow vacancy left between the boards of one-third of an inch wide, near the top, to let out the foul air; and perhaps it may be neceffary to nail along the upper edge of thefe openings lift, or flips of fail-cloth, to hang over them, to fecure the plants from any fpray of the fea; and at the fame time it will not prevent the air from paffing through. Boxes with plants packed in this manner, muft be placed where there is free air, that is, out of the way of the foul air of the fhip's hold.

If the plants fhould be taller than the depth of the box men-tioned here, they may be placed length-ways in the fame fized boxes: but then care muft be taken to fecure their roots in the

mofs

mofs at one end of the box, fo as not to be fhook out of their places, and laths fhould be nailed acrofs the infide to fupport their branches, and keep them from preffing upon one another. The beft mofs that can be ufed on thefe occafions is the *Sphagnum paluftre*, or fwamp mofs, which is very foft, whitifh, and fpongy; it will retain water a long time, and not be liable to putrefy.

The following method of preferving feeds from turning rancid from their long confinement, and the great heat of the climates which they muft neceffarily pafs through from China, was communicated to me fome years ago by the celebrated Profeffor Linnæus, of Upfal, in Sweden. He advifes, that each fort of feed fhould be put up in feparate papers, with fine fand among them, to abforb any moifture (dried, loamy, or foapy earth may be tried): thefe papers, he fays, fhould be packed clofe in cylindrical glafs, or earthen veffels, and the mouths covered over with a bladder, or leather tied faft round the rims: he then directs that thefe veffels, with the feeds in them, fhould be put into other veffels, which fhould be fo large, that the inner veffel may be covered on all fides, for the fpace of two inches, with the following mixture of falts. Half common culinary falt; the other half to confift of two parts of falt-petre, and one part of fal-ammoniac, both reduced to a powder, and all thoroughly mixed together, to be placed about the inner veffel, rather moift than dry. This he calls a refrigeratory; and fays it will keep the feeds cool, and hinder putrefaction. Perhaps if fmall tight boxes, or cafks or bottles of feeds were inclofed in cafks full of falts, it might be of the fame ufe, provided the falts do not get at the feeds; and as fal-ammoniac may not be eafily met with, half common falt, and the other half falt-petre, or common falt alone, might anfwer the fame end. But it would be very neceffary to try both methods, to know whether the latter would anfwer the purpofe of the former, as it would be attended with much lefs trouble, and might prove a ufeful

C 2

method

method to our feedfmen, in fending feeds from hence to thofe warm climates.

The fmalleft feeds being very liable to lofe their vegetative power by long voyages through warm climates, it may be worth while to try the following experiment upon fuch kinds as we know for certain are found. Dip fome fquare pieces of cotton cloth in melted wax, and while it is foft and almoft cold, ftrew the furface of each piece over with each fort of fmall feed, then roll them up tight, and inclofe each roll in fome foft bees-wax, wrapping up each of them in a piece of paper, with the name of the feed on it; thefe may be either furrounded as before with falts, or packed without the falts in a box, as is moft convenient.

There are many feeds, which we receive both from the Weft-Indies and the fouthern parts of our North-American colonies, as South-Carolina, Georgia, &c. which the gardeners find very difficult to raife here, unlefs the following method is purfued. Divide a box, according to your quantity and forts of feeds, into feveral fquare partitions ; then mix the feeds with loamy earth and cut mofs, and put each fort into its feparate cell, filling it up to the top: the earth and mofs muft be rather inclining to dry than wet; then nail the lid down very clofe on your box, keeping it in an airy fituation. If the voyage does not exceed two months, they will arrive in good order in the fpring; and, though many of them may begin to germinate, yet, if they are fown directly, they will fucceed much better than thofe that are brought over in papers, as is well known to our moft curious gardeners. Seeds of the nutmeg-tree from Tobago, the cinnamon-tree, the cacao or chocolate-nut, and Avocado pear, muft be brought in this manner. Seeds of all the forts of magnolias, ftewartias, chionanthus, and many others from South-Carolina, will fucceed better this way, than any other method we yet know.

There

There are many valuable seeds may be brought from the South of France, Italy and Turkey, particularly the rarer kinds of oaks, the Alkermes oak, the Velani oak, the gall-bearing oak, which ought to be preserved in bees-wax, as the voyage is often very long, and the Turkey ships frequently detained on account of the quarantine.

The seeds of many of the small succulent fruits may be brought to England from very distant parts, by pressing them together, squeezing out their watery juices, and drying them in small cakes gradually, that they may become hard; they may be then wrapt up in white writing paper, not spongy, as this is apt to attract and retain moisture: but I believe it will be found, that a covering of wax will be better than one of paper.

The Alpine strawberry was first sent to England in a letter from Turin to Henry Baker, Esq; F. R. S. by pressing the pulp with the seeds thin upon paper, and letting it dry before they were inclosed. The paper mulberry from China was brought hither about the year 1754, much in the same manner. Formerly, varieties of the Arbutus, from the southern parts of France, were brought over in thin dried cakes; and a few years ago the Arbutus Adrachne seeds were sent in the same manner from Aleppo by the late Dr. Al. Russel. Our mulberries, strawberries, and other succulent fruits, may be conveyed to distant parts by the same method. The pulp, when dried, hardens like a varnish, and keeps the seeds from the air (provided they are kept dry), as the larger kinds are by bees-wax.

These hints may prompt us to try the larger succulent fruits; for instance, the mangoes, lechees, and others of this kind: if their fleshy part, when they are very ripe, was brought to the consistence of raisins or dried figs, it would keep their kernels plump, and in this state they might be better preserved in wax, than by any other method yet known. The nutmegs in the same manner must not be divested of their pericarpium before they are inclosed in wax. The marking nut, or anacardium orientale, should be brought over

with

with its apple or receptacle dried, adhering to it before it is in-closed in wax. Of this valuable plant we are yet ignorant, even of its leaves and blossoms, though very East-India ship brings some of the nuts, but none of them have yet been raised in England. This is the tree so much commended by Kæmpfer, in his Amœni-tates Exoticæ, p. 793. for yielding the Siam varnish of so much consequence in China and Japan, for the first layer of their varnish, in all their curious lacquered ware. There is another fruit which I shall recommend to be brought in wax from China; this is called by the Chinese Unchee, or Um-Ky; it is described by Doctor Solander, in the Philosophical Transactions, vol. LII. p. 654. Feb. 20. where there is a very exact figure of it, taken from spe-cimens in the British Museum, as they are preserved in several Hortus Siccus's; the volumes in which they are to be found are particularly enumerated by the Doctor, with an account of their great use in dying scarlet: this shrub may be cultivated in our American islands. The pulpy part among the seeds gives, when put into warm water, a very lively yellow colour, which is much wanted among the dyers. This plant is now cultivated in our curious botanical gardens from cuttings, and is known by the name of the single Gardenia, or the single Cape Jasmine of Miller: it was raised from seeds about ten years ago, brought from China by Thomas Fitzhugh, Esq; and is often found among the boxes of seeds sent from China, but not in a vegetating state. Mr. Fitzhugh followed the Linnæan manner of bringing over seeds surrounded with salt, which he thinks a very good method.

Our seedsmen are much distressed for a proper method to keep their seeds sound, and in a state of vegetation, through long voyages. Complaints are made, that, when their seeds arrive in the East-Indies, and often in the West-Indies, few of them grow; but that most of them are full of insects, or, what they term weevilly. This seems to proceed from the damp and putrid heat

of

of the hold, or too long confinement in close warm air, which brings these animals to life, which soon begin to prey on the inside of the seeds, and those seeds that are oily turn rancid. This putrid penetrating steam, that strikes every one upon opening the hatches of a full loaded ship's hold after a long voyage, it is this that does the mischief to seeds. This vapour, as the excellent Doctor Hales observes, without frequent ventilating, will become fatal to vegetable substances, as well as animals.

When the cavalry of our army in Germany was under the necessity of being supplied with hay from England, the difference was too manifest between the hay that had been but a month on board, and fresh hay, that had never been confined in the hold of a ship.

Experiments have been made on the best hemp from Russia, and hemp of English growth, by persons belonging to the navy, of great credit and honour, and the difference in the strength was amazing; the length of the voyage from Russia, with the very close package that is necessary to stow that article on board of a ship, raises such a heat, as to shew evident signs of putrefaction begun, which must weaken the strongest vegetable fibres*.

To illustrate this farther in an instance of the different manner of packing and stowing seeds for a long voyage, which has lately come to my knowledge and may be of use, as it not only points out the error, but in some manner how to avoid it.

A gentleman, going to Bencoulen in the island of Sumatra, had a mind to furnish himself with an assortment of seeds for a kitchen

---

* This hint may be worthy of the consideration of the linen as well as the hempen manufacturers, both in Great-Britain and Ireland, as it will shew them the necessity of raising both hemp and flax, the first principles of these most useful and necessary manufactures, at home; by convincing them, from experiment, of the great difference they will find between the comparative strength of what we raise at home, and what we bring from beyond sea.

garden;

garden; thefe were accordingly packed up in boxes and cafks, and ftowed with other goods in the hold of the fhip.

When he arrived at Bencoulen, he fowed his feeds; but foon found, to his great mortification, that they were all fpoiled, for none of them came up.

Convinced, that it muft be owing to the heat of the fhip's hold, and their long confinement in putrid air, and having foon occa-fion to return to England; he determined in his next voyage thi-ther, to pack them up in fuch a manner, and place them fo, as to give them as much air as he could, without the danger of expofing them to the falt-water; and therefore put the fmaller kinds into feparate papers, and placed them among fome clean ftraw in a fmall clofe net, and hung it up in his cabbin; and the larger ones he put into boxes, ftowing them where the free air could come at them, and blow through them: the effect was, that as foon as he arrived at Bencoulen he fowed them, and in a little time found, to his great fatisfaction, that they all grew extremely well. It is well known to our feedfmen, that, even here at home, feeds kept in clofe warehoufes, and laid up in heaps, frequently fpoil, unlefs they are often fifted, and expofed to the air.

Seeds faved in moift cold fummers, as their juices are too watery, and the fubftance of their kernels not fufficiently hardened to a due ripenefs, are by no means fit for exportation to warmer climates.

Our acorns, unlefs ripened by a warm fummer, will not keep long in England: thofe acorns that are brought from America, and arrive early in the year, generally come in good order, owing to their juices being better concocted by the heat of their fum-mers, and are not fo apt to fhrivel when expofed to the air as ours are.

Thefe

These hints are given to shew how necessary it is to take care, that the seeds we send abroad should be perfectly ripe and dry.

One of the methods now practising in sending garden-seeds to the East and West-Indies, is to put a small piece of camphire into each parcel: as to this experiment, we are not yet certain of its success; the hint is taken from the common method of preserving butterflies, moths, beetles, and other insects, from being destroyed by very minute animalcules, which are apt to infest them.

Flowers of sulphur in water, in a certain proportion, will destroy insects that infest plants, and will rather encourage than hurt their vegetation, as appears from a method practised here with success for many years, in the culture of the ananas, or pineapple plant, by one of the most eminent fruit and kitchen gardeners in England *. The inside of boxes and casks should be washed

with

* In order to introduce the method of destroying insects that infest the ananas, or pine-apple plant, it may not be disagreeable to the reader, to know some general rules (though foreign to our present subject), that are necessary to be observed in the culture of that curious and delicate fruit.

" The stems of the heads and suckers should not be stripped up higher than to the place
" where they appear white under the leaves you pull off.

" The composition to plant them in should be three parts of strong fresh loam, and the
" fourth part rotten dung; they should be mixed together, and often turned, for a year at least
" together before it is used. The pots should be rather small than large, in proportion to the
" plants at all times.—The plants should be put into the stove or store-pit, and kept with a
" brisk heat, shading them from the violence of the sun, and sprinkled every day, or twice a
" day, if the weather is hot.—In a week they will have roots enough to support themselves,
" and should be inured by degrees to the full sun, and the oftener they are sprinkled in warm
" weather, the faster they will grow; but when they are sprinkled, they should be shut up
" close, and shaded for an hour or more; then give them air, and take away the shade.
" Those plants that are large, and that you design should bear fruit the next year, should be
" put into larger pots the latter-end of August, when some new tan should be added, and
" mixed up with that which they stood in till this time.—In November, the tan-bed should
" be turned over two-thirds of the way down, and a good quantity of new tan mixed with it,
" throwing away some of the rottenest, which may be separated by screening it: this will heat
" sufficiently to carry the plants on till January or February, when they will shew their fruit,
" if the fire heat is kept up as usual. As soon as the plants begin to grow in the spring,

D

" they

with water that has been impregnated with fulphur; or, perhaps, the Hepar Sulphuris, or liver of fulphur, which is fulphur combined with an alkaline falt to make it foluble in water, would be more effectual: a little of this folution laid over the infide of a box or cafk, with a hog's-briftle brufh, would raife fuch a penetrating ftench in warm weather, when confined, as to deftroy all kind of infects. Or the cafks and boxes might have brimftone burnt in them before the feeds are put in them: but the fuccefs muft depend on experiment. There is great probability, that the vegetative powers of the feeds will not be hurt by the fumes of the fulphur, if we may reafon from the ufe of it in deftroying the infects in the pine-apple, and rather promoting than hurting their vegetation†.

Befides

" they fhould be often fprinkled with water made a little warm only, by ftanding in the
" ftove. But when the plants are in bloom, care muft be taken not to wet the bloffom,
" which would prevent the fruit fwelling near fo large as it would if they had been kept dry
" at that time.
" In February or March, before the plants blow, the tan-bed fhould be turned over, and a
" little more new tan added to it, and all the plants that have fhewed for fruit fhould be put
" into larger pots; but not to put any plants that you defign to have fruit into larger pots,
" till they fhew for fruit, nor fhould any of their roots be cut off; but take off all the earth,
" from the furface down to the roots, of thofe you put into larger pots. The fooner you
" fhift your ftove plants in the fpring into frefh earth and larger pots the better, as it will be
" a means, not only of fetting them a-growing early, but keeping them from fruiting. It is
" a practice among our nurfery-men, to force the young plants in hot-beds of horfe-dung
" with a moift ftrong heat, which pufhes them forwarder than tan-beds during their growing
" ftate, which is from March till the end of September.
" Left the tan in the fruiting ftove fhould cool fuddenly, either through neglect or want
" of judgement, it would not be improper to have a flue run zig-zag under the bottom of the
" tan-pit, the top of which fhould be level with the bottom of the tan-pit, but not to be made
" ufe of on any account, unlefs when the heat fuddenly leaves the tan.
" If the plants are troubled with infects, take a pound of flower of brimftone and put it into
" ten gallons of water, and water the plants well all over with it. This will deftroy the in-
" fects, and promote vegetation."——It muft be remembered, that the water muft be of the
fame degree of warmth with the air in the ftove.
† Various kinds of pulfe and grain, which I have lately received from different parts of the
Eaft-Indies, have been eaten hollow, and moft of them deftroyed, by a kind of very fmall
beetle, or infect of the weevil kind.

It

Besides this method of deſtroying inſects, there is another, which, for the benefit of mankind, ſhould be generally known, particularly as moſt ſhips that paſs through warm climates are infeſted with thoſe diſagreeable ones called cock-roaches.

The following preparation will prevent them from ſpoiling many valuable articles on the voyage, and perhaps be of uſe in ſaving ſeeds, books, and papers, which they are apt to deſtroy on board of ſhips: at the ſame time we ſhall find that this preparation is equally deſtructive to all other inſects. Diſſolve one ounce of crude ſal-ammoniac in a quart of water, then put in two onnces of corroſive ſublimate mercury. This ſolution, when uſed, ſhould be firſt heated in the following manner: put the liquor into a phial, and ſet it in a tin pot of water on the fire, and when the water boils, the ſolution in the phial will be heated enough. N. B. The phial with the ſolution muſt be put into the water when it is cold, and then there will be no danger of breaking the phial: a piece of packthread or wire ſhould be put round the neck of the phial, to lift it in and out of the water: it will corrode every veſſel but glaſs; therefore it is neceſſary that it ſhould be heated in the foregoing manner. You muſt uſe a hog's-briſtle bruſh to waſh over any box or furniture.

It is likewiſe too well known, the great damage done to wheat by this pernicious inſect the weevil, which, after feeding on the inſide of the grain, eats its way out: how it comes into the grain, is a conſideration worthy of the attention of the philoſopher.

The ſame obſervation may be made on turnip-ſeeds kept confined in ſacks in hot weather, where the moiſt heat brings the animals to life. This affords us a very uſeful hint in regard to the black fly, as it is called, that deſtroys the newly-ſown turnips in dry weather, juſt as their lobe leaves are expanded; and points out to us the probability that this little animal, which afterwards deſtroys the tender plant, may have exiſted in the ſeed itſelf; ſo that it is a conſideration well worth the farmer's notice, to try either by fumigating his ſeed well with burnnig brimſtone, or by ſoaking them in brimſtone and water, or by diſſolving a ſmall portion of liver of ſulphur in water, and ſteeping his ſeed in it, to deſtroy theſe animals. If theſe experiments are made with judgement, there is great probability that we ſhall able to deſtroy the animals without affecting the vegetation of the plant.

The

The heating of the liquor will make it penetrate better into wood, and no insect will come near where it has been once rubbed over. If this liquor is put into the paste used in binding of books, the cockroaches or other insects will never touch them. It will likewise preserve the hair and feathers of dried specimens of birds and beasts, and the bodies of curious butterflies, from being destroyed by minute animalcules; and will be found to be an effectual remedy against bugs, and is one of the great secrets of the bug-killers. Another is a solution of corrosive sublimate mercury, dissolved in spirit of wine, and lowered with water: this leaves no stain on furniture.

As tobacco is universally known by the gardeners to destroy insects by its deleterious quality, and as tobacco-sand is to be had upon very reasonable terms, it is recommended to seedsmen to mix it up with their smaller seeds on exportation, as it may absorb their humidity, prevent their putrefaction, and destroy the insects that are in them. But it must be observed, that it is not meant here, that it will keep them from the penetrating noxious steams that arise from the ship's hold, particularly in warm climates; for I am doubtful, whether even a thin coat of wax would be a sufficient guard in that dangerous situation. And as to the larger seeds, the putting some fine cut-tobacco in small quantities loose among them, seems to carry some probability of being at least an experiment worth trying, to prevent their being destroyed by insects.

In short, the demand for our kitchen-garden-seeds would be very great, both in the East and West-Indies, if we could hit on a proper method of sending them into those warm climates in a vegetating state; so that it is well worth our attention, as an article of commerce, to try every experiment that may lead to so useful a discovery.

I hope, then, these hints may incite curious gentlemen, as well as intelligent seedsmen and gardeners, to begin a course of these
kind

kind of experiments; in the progreſs of which, I am perſuaded, they will receive great pleaſure as well as knowledge, and both do honour to themſelves, and a real ſervice to their country. But as gardeners and ſeedſmen, from their conſtant experience, muſt know the nature of ſuch bodies better than moſt gentlemen, eſpecially as it is their daily buſineſs; I don't doubt but that excellent and uſeful Society for the encouragement of arts, manufactures, and commerce, will amply reward their diſcoveries.

⁂ It may be neceſſary to add to the article of preſerving ſeeds in wax, that whereas many of the valuable kinds, ſuch as cloves, pepper, &c. are too ſmall to be rolled up in wax ſeparately; many of them may be incloſed in ſmall balls of warm wax in ſuch a manner, as to be kept from touching each other; and when the balls are cold, they may be put into melted wax, in the ſame manner as in the experiment to preſerve oak acorns, tea-ſeeds, &c. in wax, before mentioned.

The

The following Catalogue of such Plants as deserve the particular Attention of our American Colonies, are here exhibited in one View, in order to incite such Persons as have it in their Power to procure the Seeds or Plants of the most valuable of them, for this interesting Purpose.

To avoid Confusion in the Botanical Names, both the generical and specific, or trivial Names of the Plants, are set down, with the Page referred to in the celebrated Linnæus's Second Edition of his Species of Plants.

Other Authors of the best Authority are mentioned, where Linnæus is silent.

Latin Names.	2d Ed. Lin. Sp.	English names.	Observations.
Rubia Peregrina Rubia Tinctorum	p. 158 p. 158	Turkey Madder Dyers Madder *	The first is supposed to be the same that is now cultivated in Smyrna for a crimson dye.
Quercus Suber	p. 1413	Cork-bearing oak	Grows in the southern parts of France, Spain, and Portugal.

* This plant is a native of the warmest parts of Europe, and is better calculated for the climate of the Floridas than either of Holland or England, where it is cultivated; but principally in the former, from whence we are chiefly supplied with this valuable dye. The chemists say, and with reason, that the warmth of the climate exalts the colour. If so, it may be well worth the attention of the publick to encourage the planting of so valuable an article of commerce in a climate and soil that seems so much better adapted to it, where the land is cheap, and where vegetation is so much quicker and more luxuriant; and while we encourage the growth of it in our colonies, we may have the advantage of manufacturing this valuable commodity at home, for which at present we pay *sums scarcely credible*, to the Dutch.

Quercus

Latin Names.	2d Ed. Lin. Sp.	English Names.	Observations.
Quercus Ægilops	p. 1414	Avellanea or Vale-nida oak	The cups of the acorns, which are very large, used here in dying, grow in Greece and Natolia, particularly in the Island of Zia in the Archipelago, where Tournefort says they gather in one year 5000 hundred weight.
Quercus Gallifera	Parkinson 1386	Gall-bearing oak	Galls from Aleppo and Smyrna. This oak is not yet known in England: The Acorns may be brought over in Wax, and sent to the Floridas, Georgia, and S. Carolina.
Carthamus Tinc-torius }	Lin. Sp. 1162	Safflower	Much used in dying, grows in Egypt.
Rhamnus catharticus minor	Tournft. 593	Buckthorns that produce yellow berries of Avignon.	Used by painters and dyers; both these plants produce berries fit for this purpose.
Rhamnus Saxatilis	Lin. Sp. 1671		
Olea Europea	p. 11	Olives of several varieties	For oil; these grow in France, Spain, and Italy. Young Plants and ripe Fruit of the French and Spanish sorts, may be brought from thence.
Sesamum Orientale	p. 883	Oily grain	Propagated in the Levant for oil, which does not soon grow rancid by keeping.
Gossypium her-baceum Gossypium hirsu-tum }	p. 975	Two sorts of annual cotton	Both these kinds of annual cotton are yearly sown in Turkey, and would grow well in the warm climates of N. America, as the Floridas, Georgia, Carolina, and Virginia.

Salsola

Latin Names.	2d Ed. Lin. Sp.	Englifh Names.	Obfervations.
Salfola Soda Salfola Sativa } and Chenopodium maritimum	p. 323 p. 321	Thefe kinds of glafswort for Barilla	Thefe are fown yearly in fields near the fea in Spain, for making Barilla, for foap, glafs, &c.
Ceratonia Siliqua	p. 1513	Locuft-tree or St. John's Bread	The pods are excellent food for hard-working cattle, and ufed for this purpofe on the fea-coaft of Spain, where they are eafily propagated from feeds or cuttings.
Piftachia Vera	p. 1454	Piftachia-tree	They are propagated about Aleppo, where the female or fruit-bearing ones are ingrafted on the ftocks raifed from the nuts.
Piftachia Terebinthus	p. 1455	Chio turpentine-tree	This kind of turpentine is ufed in medicine.
Piftachia Lentifcus	p. 1455	Maftick-tree	Gum Maftick from the ifle of Scio; as this tree, commonly called the Lentifcus, is doubted to be the genuine Maftick-tree, feeds of the true kind may be procured from the ifle of Scio.
* Styrax Officinale	p. 635	Gum Storax-tree	This tree grows in Italy, Syria, and India; but the warmer climates yield the beft gum.
Convolvulus Scammonia	p. 218	Gum Scammony	Seeds of the Plant, from whence this excellent drug

* There is a refinous juice, which, by age, hardens into a folid brittle refin, of a pungent, warm, balfamic tafte, and very fragrant fmell, not unlike the Storax calamita, heightened with a little ambergrife, which is produced from the Styrax aceris folio of Ray, or Liquidambar Styraciflua of Linnæus, Spec. plant. 1418, which grows in perfection in the Floridas. This, Dr. Lewis, in his Materia Medica, p. 353, fays, might be applied to valuable medicinal purpofes. The French, in Du Pratz Hiftory of Louifiana, fpeak with rapture of its healing qualities, and the high efteem it is in among the Indians of Florida, on account of its infinite virtues: it is known to the Englifh by the name of the Sweet Gum-tree, and to the French by the name of Copalm.—This is well worth the attention of the College of Phyficians, as we can have it genuine, whereas the Storax from the Eaft is often adulterated.

Latin Names.	2d Ed. Lin. Sp.	Englifh Names.	Obfervations.
			is procured, were fent into England about 20 Years ago, from Aleppo, by the late Dr. Alex. Ruffel: it bears this climate very well, and produces feed in hot fummers; but requires the warmer climates of Carolina, Georgia, and the Floridas, to make the gum-refin that flows from it a beneficial article of commerce. It is fo frequently adulterated in Turky, that, in order to have it genuine, it is well worth propagating in our colonies.
Papaver Somniferum	p. 726	True opium poppy	This is recommended to be fown in our fouthern colonies of North-America, for the fake of obtaining the opium pure*.
Caffia Senna	p. 539	Alexandrian purging Senna	This grows in Upper Egypt, and is brought from thence to Alexandria; it would not be difficult to procure the feeds of this ufeful drug.
Croton Sebiferum	p. 1425	Tallow-tree of China	This plant grows in moift places in China, and is of great ufe in that country.
Rheum Palmatum	p. 521	True Rhubarb	The feed of this plant was brought to England about five years ago, by Dr. Mounfey, F. R. S. from Mofcow, and appears by experiment to be the genuine true Rhubarb of the fhops, and is a

* The feed of this fpecies of poppy is recommended by a phyfician of great eminence as proper for the fame purpofes in medicine as fweet almonds are ufed. It is obferved not to have the leaft degree of a narcotic quality in it.

E

moft

Latin Names.	2d Ed. Lin. Sp.	English Names.	Observations.
			most valuable acquisition to this country, as it will grow well in a deep rich soil, inclining to a sandy or gravelly loam, but not in too wet a situation, and may be cultivated both here and in North-America. Mr. Inglish has raised this plant with so much success at his country-house at Hampstead, as to be able not only to produce some excellent good Rhubarb, but a sufficient quantity of ripe seed to make a large plantation; and at the same time has most generously bestowed a great deal of seed to be sent to our American colonies, where, no doubt, but it will prove in a few years a most beneficial article of commerce.
Calamus Rotang	p. 463	Three sorts of Gum Dragon, or Dragon's blood	1. From a kind of cane in the East-Indies. 2. From Java and Surinam. 3. From the Canary and Madeira islands.
Pterocarpus Draco	p. 1662		
Dracæna Draco	Lin. Syst. Ed. 12. p. 246		
Dolichos Soja	Lin. Sp. 1023	A kind of kidney-bean called Daid-su	Used for making Soye † or India Ketchup. See Kæmp. Amœnitat. 837.

Laurus

† The method of preparing East-India Soye, or India Ketchup.
Take a certain measure, for instance a gallon, of that sort of kidney-beans, called Daidsu by the Japonese, and Caravances by the Europeans; let them be boiled till they are soft; also a gallon of bruised wheat or barley, (but wheat makes the blackest Soye) and a gallon of common salt. Let the boiled caravances be mixed with the bruised wheat, and be kept covered close a day and a night in a warm place, that it may ferment. Then put the mixture of the caravances and wheat, together with the gallon of salt, into an earthen vessel, with two gallons and a half of common water, and cover it up very close. The next day stir it about well with a battering machine or mill (Rutabulum) for several days, twice or thrice a day, in order to blend it more thoroughly together. This work must be continued for two or three months, then strain off and press out the liquor, and keep it for use in wooden vessels; the older it is the clearer it will be, and of so much more value. After it is pressed out, you may pour on the remaining mass more water, then stir it about violently, and in some days after you may press out more Soye.

Latin Names.	2d Ed. Lin. Sp.	Englifh Names.	Obfervations.
Laurus Caffia	p. 528	Caffia Lignea-tree	Grows in Sumatra.
Laur. Cinamomum	p. 528	Cinnamon-tree	In Ceylon, Guadaloupe, and in moft of our newly ceded iflands.
Laurus Camphora	p. 528	Camphire-tree*	In Japan, and in Sumatra, now in England in the green-houfes about London. It will grow freely where oranges and lemons do.
Cycas Circinalis	p. 1658	Sago Palm-tree	In Java, and the warmeft parts of the Eaft-Indies.
Amyris Gileadenfis	Lin. Mant. 165.	True balm of Gilead-tree †	Lately difcovered in Arabia by Dr. Forfkall, and defcribed by Dr. Linnæus in a late differtation.
Arundo Bambo	p. 120	The true Bamboo cane	Of great ufe in China, and might be alfo in our American iflands ‡.
Anacardus Orientalis	Kæmp. Amœn. p. 793	Siam varnifh-tree, called Ton-rak by the Japonefe	The fruit of this is the Malacca bean, or marking nut, and the Oriental Anacardium of the fhops. This is the common varnifh of the Eaft-Indies, as defcribed by Kæmpfer. This tree is unknown to the botanifts.

* The camphire from Sumatra is greatly preferable to that of Japan; we are not certain whether it is from a different fpecies of tree, but it feems well worth inquiring into as the effects of proportionable quantities in medicine are furprizingly different, perhaps it may be owing to the great difference of heat in the climates.

† We have in the ifland of Jamaica, a fpecies of tree of this genus, called by Linnæus Amyris balfamifera. See Species Plantarum, p. 496. Sir Hans Sloane, in his Hift. of Jam. vol. II. p. 24. calls this tree Lignum Rhodium, from the odoriferous fmell of its wood when burnt, which it diffufes a great way; for which reafon he believes it to be the tree that afforded the agreeable fcent which Columbus perceived on the fouth fhore of Cuba, upon the difcovery of that ifland, as it is mentioned by feveral hiftorians.—Dr. Pat. Browne, in his hiftory of Jamaica, p. 208. calls this tree white candlewood, or rofewood, and commends it much; he fays it is very refinous, burns freely, and affords a moft agreeable fmell; and that all the parts of this tree are full of warm and aromatic particles.——Quere, Whether it is not worth while to extract the balfam, as it agrees fo near in character and genus with that moft valuable drug the balfam. of Mecca?

1 The French had brought this moft ufeful plant from the Eaft-Indies to their Weft-India iflands: a few roots have been got from thence to Grenada, and will perhaps in time become

familiar

Latin Names.	2d Ed. Lin. Sp.	English Names.	Observations.
Thea	Lin. Sp. p. 734	Tea	From Japan and China. See Kæmpfer's Amœnitates, p. 60 *.
Gardenia Florida.	p. 305	Umky of the Chinese	Used in dying scarlet in China. The pulp that surrounds the seeds, gives in warm water a most excellent yellow colour, inclining to orange. See Phil. Transf. Vol. 52. p. 654. where there is an exact figure of it.

familiar in our islands. But too much pains cannot be taken in the propagation of this plant as its uses are manifold and extensive, both in building, and all kinds of domestic instruments.

* It is asserted by some people, that the green tea and the bohea tea are two different species; but without foundation: they are one and the same species. It is the nature of the soil, the culture, and manner of gathering and drying the leaves, that makes the difference; for take a green tea-tree and plant it in the bohea country, and it will produce bohea tea, and so the contrary. This is a fact attested by gentlemen now in London, that have resided many years in China, and who have had great experience in this article.

The method of bringing over this valuable plant being already described, I shall only mention an observation of the celebrated Linnæus, who is now in possession of the true tea-tree, two of which he received from Captain Ekenberg, the commander of a Swedish East-Indiaman; in the year 1763, who raised them from seed on the voyage. This celebrated professor had tried for many years to get this curious tree into the physic-garden at Upsal; but, by a variety of accidents, they were all destroyed on the passage. At length, about the year 1755, Mr. Lagerstrom, a director of the Swedish East-India Company, brought him two plants alive to the garden at Upsal, wich he had bought in China: they grew very vigorously for two years; but when they came to shew their blossom, they proved to be of that genus of plants, called by Kæmpfer Tsubakki, and by himself Camellia, Sp. Plant. p. 982. The crafty Chinese, when they sold the plants to Mr. Lagerstrom for the true tea plants, had artfully pulled off the blossoms.

Kæmpfer observes, that there is one species of Tsubakki, (see his Amœnit. Exoticæ, p. 853. the leaves of which they prepare, and mix with their tea, to give it a fine flavour; and Linnæus says, that the leaves of his Camellia are so like the true tea, that they would deceive the most skilful botanist: the only difference is, that they are a little broader. In a letter, dated Upsal, November 8, 1769, he says, that he has just received from a very great person in France, a small branch of a plant, which was brought from China for the true tea; but it proves to be the Camellia. This caution is intended for captains of East-India ships, not to purchase the plants, but the fresh seeds of the tea in their capsules; which they may soon sow after they pass the Cape of Good Hope, or on other parts of the voyage as directed.

A new kind of tea-tree being this last summer brought from China, it is suspected may be a Camellia; but as that is a most elegant flowering shrub, it may be as valuable an acquisition to the gardeners as a tea-plant, considering the many tea-seeds that have succeeded lately, which have been brought home in wax, and otherways.

The late Lord Petre, of Thorndon-hall in Essex, was formerly in possession of one of these beautiful Tsubakki's, or Camellia's, which was greatly admired for the elegant brightness of its flowers. See the figure in Edwards's History of Birds, vol. ii. t 67.

This

Latin Names.	2d Ed. Lin. Sp.	English Names.	Observations.
Mangifera Indica	p. 290	East-India Mango-tree	This excellent fruit is much esteemed in the East-Indies, and 'tis said there is a tree of it now growing in the island of Madeira. By the description which Dr. Solander gives of this fruit, at Rio Janeiro in Brasil, it is not so good as the East-India sort.
Morus papyrifera	p. 1399	Paper Mulberry-tree	Used for making paper in China and Japan. See Kæmp. Amœnit. p. 467. This has been some time in the English gardens.
Cinchona Officinalis	p. 244	Jesuits-bark tree	This grows at Loxa in the province of Peru; and could it be obtained so as to be cultivated in our American islands, would be of infinite advantage to us.
Dorstenia Contrayerva	p. 176	Contrayerva-root	This grows in New Spain, Mexico, and Peru.
Smilax Sarsaparilla	p. 1459	Sarsaparilla-root	It is brought from the Bay of Campeachy, and the Gulph of Honduras, where it grows in plenty, and might easily be propagated in Florida.
Copaifera Officinalis	p. 557	Balsam Copaiva tree	In Brazil, and Martinico.
Toluifera Balsamum	p. 549	Balsam Tolu tree	This tree grows near Carthagena, in South-America.
Hymenea Courbaril	p. 537	The Locust or Gum Copal tree, for the finest transparent varnish.	This tree is known to yield the true Gum Copal, and that the difference between this and Gum Anime, may be

Latin Names.	2d Ed. Lin. Sp.	English names.	Observations.
			be owing to soil and heat of climate; it grows wild in our American islands, the Moskito shore, and in Terra Firma.
Jalapium Officinarum	Dale 183	True Jalap	This plant is supposed by some to be a kind of Bindweed or Convolvulus, that grows near Mexico; by others it is thought to be a species of Marvel of Peru. As we are uncertain of the genus it is well worth enquiring into, as a most useful drug, in order to propagate it in our colonies.
Bixa Orellana	Lin. Sp. 730	Arnotto, for dying	This grows in all the warm climates of America. The French cultivate it, but what the Spaniards send is much richer in colour and more valuable.
Mimosa Senegal	p. 1506	Gum Senegal tree	This grows in Ægypt, and in Senegal.
Mimosa Nilotica	p. 1506	Gum Arabick	In Ægypt, from whence the seeds may be procured.
Ficus Sycomorus	p. 1513	True Sycamore of Zaccheus	This is reckoned the most durable timber we know. The repositories of the Mummies found in Ægypt are made of this timber.
Ficus Carica	p. 1513	Turkey Figs	Figs grow in the greatest perfection in Carolina, and would become a valuable trade if they had the method of curing them as in Turkey.

The

Latin Names.	2d Ed. Lin. Sp.	English Names.	Observations.
Vitis Apyrena	p. 293	Currants or Corinthian grapes	The cuttings of this vine might be procured from Zant.
Fraxinus Ornus	p. 1510	Calabrian Manna Ash *	This is worth trying in our southern colonies, where the heats are violent in the summer. It is common in our nursery gardens.
Amygdalus Communis	p. 677	Sweet Almonds	These would grow to great perfection in our southern colonies.
Capparis Spinosa	p. 720	Caper tree	This shrub requires a rocky soil to grow in, as it is about Marseilles and Toulon.
Punica Granatum †	p. 676	Balaustians, or the blossoms of the double flowering pomegranate	This tree would thrive extremely well in our southern provinces, and yield a profitable article in their blossoms. Plants of this kind are to be bought from most of our nursery-men.
Lichen Roccella	p. 1622	Argal, Canary-weed, or Orchell	Tis possible this valuable plant may be found in our American islands, as well as in the Canaries and Cape-Verd islands.
Cistus Ladanifera	p. 737	Gum Labdanum	In Spain and the Archipelago.
Bubon Galbanum	p. 364	Gum Galbanum	In Ethiopia.

* There is no drug so liable to adulteration as this: and therefore, as it is a medicine so frequently in use among persons of tender constitutions, especially young children, great care should be taken to have it genuine.

† The single flowering or fruit-bearing Pomegranate, will afford the most grateful addition to the fruits of our colonies, and a valuable medicine. The ripe fruit full of seeds is to be met with at our fruit-shops in the winter season: from the seeds of such fruit this tree may be easily propagated.

Pastinaca

Latin Names.	2d Ed.Lin. Sp.	Englifh Names.	Obfervations.
PaftinacaOpoponax	p. 376	Gum Opoponax	In Sicily.
Amomum Carda- momum	p. 2	Cardamums	In the Eaft-Indies.
Curcuma Longa	p. 3	Tumerick	In the Eaft-Indies.
Aftragalus Traga- cantha	p. 1073	Gum Tragacanth or Gum Dragon	In the fouth of France and in Sicily.
Cucumis Colycin- this	p. 1435	Coloquintida, or Bitter apple	In Africa
Gentiana lutea	p. 329	Gentian	In the Alps, Apennines, and Pyrenees. To be had of the nurfery-men.
Similax China	p. 1459	China root	In China and in New Spain.
Pimpinella Anifum	p. 379	Anife feeds	In Egypt.
Gambogia Gutta	p. 728	Gamboge	In the Eaft Indies.
Quercus Coccifera	p. 1413	Alkermes oak	About Marfeilles and Toulon.
Myrrha Offic.	Dale. 325	Gum Myrrh	In Abyffinia. The characters of this plant and the five fol- lowing are not yet known to the botanifts.
Benzionum Offic.	Dale. 303.	Gum Benjamin	In Sumatra and Java.
AmmoniacumOffic.	Dale. 119	GumAmmoniacum	In Africa
Balfamum Perua- num	Dale. 337	Natural Balfam of Peru	In Peru.
Olibanum Thus Mafculum	Dale. 348	Frankincenfe	In the Upper Egypt and in- terior parts of Africa.
Nux Mofchata Offic.	Dale. 302	Nutmegs with Mace *	In Amboyna. In

* Specimens of the Nutmeg-tree in fruit from the ifland of Tobago have been lately received by the Earl of Hillfborough, which his Lordfhip has fent, with fpecimens of many other curious plants,

Latin Names.	2d Ed. Lin. Sp.	Englifh Names.	Obfervations.
Caryophyllus aro-maticus	Lin. Sp. 735	Cloves	In the Molucca iflands.
Piper Nigrum	p. 40	Pepper	Sumatra.
Garcinia Monga-ftona	p. 635	Mangofteens	A moft delicious fruit, grows in Java, and in feveral parts of the Eaft-Indies.
Lechee		Lechee of China	This fruit is highly commend-ed by all perfons who have been in China *.
Ipecacuanha	Dale. 170 Margrave 17	Ipecacuanha of the fhops, or Brafilian root.	Very ufeful in medicine, and worthy of our attention to propagate it in our Weft-India iflands: at prefent its genus is unknown to the botanifts.
Ferula Affa Fœtida	Lin. Sp. 356	Affa Fœtida, or De-vil's dung, called Hing in the Malay language	The gum of this plant is much much ufed in medicine. Kæmpf. 535 and 536.

To this catalogue may be added liquorice, faffron, and aloes foco-torina: of the two firft we do not raife near a fufficiency at home for our own confumption, but are obliged to import thofe articles from Spain.

plants, for the information of the publick, to the Britifh Mufeum. They are certainly of the fame genus with the true nutmeg, and poffibly may be improved by cultivation; the mace evi-dently covers them, and they have all the characters and the fame leaves with the wild Nutmeg. tree defcribed by Rumphius, in his Herbarium Amboinenfe, publifhed by Burman.

* The characters of this fruit are not yet known to the botanifts.

F   A BO-

Venus's Fly-trap.

Dionæa Muscipula

A sensitive Plant from the Swamps of North America with a spike of white blossoms like
the English Ladysmock.
Each leaf is a miniature figure of a Rat trap with teeth; closing on every fly or other
insect, that creeps between its lobes, and squeezing it to Death.

James Roberts sculp.

A

BOTANICAL DESCRIPTION
OF THE

DIONÆA MUSCIPULA,
O R

VENUS's FLY-TRAP.

A NEWLY-DISCOVERED SENSITIVE PLANT:

In a LETTER to Sir CHARLES LINNÆUS,

Knight of the Polar Star, Phyſician to the King of Sweden, and Member of
moſt of the Learned Societies of Europe,

From JOHN ELLIS, Fellow of the ROYAL SOCIETIES of
LONDON and UPSAL.

F 2

London, Sept. 23, 1769.

My dear Friend,

I KNOW that every difcovery in nature is a treat to you; but in this you will have a feaft.

You have feen the Mimofa, or Senfitive Plants, clofe their leaves, and bend their joints, upon the leaft touch: and this has aftonifhed you; but no end or defign of nature has yet appeared to you from thefe furprizing motions: they foon recover themfelves again, and their leaves are expanded as before.

But the plant, of which I now inclofe you an exact figure, with a fpecimen of its leaves and bloffoms, fhews, that nature may have fome view towards its *nourifhment*, in forming the upper joint of its leaf like a *machine* to catch food: upon the middle of this lies the bait for the unhappy infect that becomes its prey. Many minute red glands, that cover its inner furface, and which perhaps difcharge fweet liquor, tempt the poor animal to tafte them: and the inftant thefe tender parts are irritated by its feet, the two lobes rife up, grafp it faft, lock the rows of fpines together, and fqueeze it to death. And, further, left the ftrong efforts for life, in the creature thus taken, fhould ferve to difengage it; three fmall erect fpines are fixed near the middle of each lobe, among the glands, that effectually put an end to all its ftruggles. Nor do the lobes ever open again, while the dead animal continues there. But it is neverthelefs certain, that the plant cannot diftinguifh an animal, from a vegetable or mineral, fubftance; for if we intro- duce a ftraw or a pin between the lobes, it will grafp it full as faft as if it was an infect.

In

In the year 1765, our late worthy friend, Mr. Peter Collinſon, ſent me a dried ſpecimen of this curious plant, which he had received from Mr. John Bartram, of Philadelphia, botaniſt to the King. The flower of this ſpecimen Doctor Solander diſſected with me, and we found it to be a new genus; but not ſuſpecting then the extraordinary ſenſitive power of its leaves, as they were withered and contracted, we concluded they approached near to the *Droſera* or *Roſa Solis*, to which they have been ſuppoſed by many perſons ſince to have a great affinity; as the leaves of the moſt common Engliſh ſpecies of *Roſa Solis* are round, concave, beſet with ſmall hairs, and full of red viſcid glands.

But we are indebted to Mr. William Young, a native of Philadelphia (to whom likewiſe the Royal favour has been extended, for his encouragement in his botanical reſearches in America), for the introduction of this curious plant alive, and in conſiderable quantities. He informs me, that they grow in ſhady wet places, and flower in July and Auguſt; that the largeſt leaves, which he has ſeen, were about three inches long, and an inch and half acroſs the lobes; and obſerves, that the glands of thoſe that were expoſed to the ſun were of a beautiful bright red colour, but thoſe in the ſhade were pale, and inclining to green.

It is now likely to become an inhabitant of the curious gardens in this country, and merits the attention of the ingenious.

The Botanical Characters of the Genus Dionæa, according to the Linnæan Sexual Syſtem, where it come under the Claſs of Decandria Monogynia.

The *Calyx,*   or Flower-cup, conſiſts of five ſmall, equal, erect leaves, of a concave oval form, pointed at the top.

The

The *Corolla*,   or Flower, has five concave petals, of an oblong, inverted-oval form, blunt at the top, which curls in at each side, and is ftreaked from the bottom upwards with feven tranfparent lines.

The *Stamina*,   or Chives, have ten equal filaments, fhorter than the petals; and their tops, which contain the male duft, are roundifh. This duft, or farina fœcundans, when highly magnified, appears like a tricoccous fruit.

The *Piftil*,   or Female Organ, has a roundifh germen or embryo feed-veffel, placed above the receptacle of the flower: this is a little depreffed, and ribbed like a melon. The ftyle is of a thread-like form, fomething fhorter than the filaments. The ftigma, or top of the ftyle, is open, and fringed round the margin.

The *Pericarpium*, or Seed-veffel, is a gibbous capfule, with one cell or apartment.

The *Seeds*   are many, very fmall, of an oval fhape, fitting on the bottom of the capfule.

I fhall now give you a general defcription of the fpecies of *Dionæa* before us, called *Mufcipula*, or *Venus's Fly-trap*.

This plant is herbaceous, and grows in the fwamps of North-Carolina, near the confines of South-Carolina, about the latitude of 35 degrees North, where the winters are fhort, and the fummers very hot.

The roots are fquamous, fending forth but few fibres, like thofe of fome bulbs; and are perennial.

2   The

The leaves are many, inclining to bend downwards, and are placed in a circular order; they are jointed and succulent: the lower joint, which is a kind of stalk, is flat, longish, two-edged, and inclining to heart-shaped. In some varieties they are serrated on the edges near the top. The upper joint consists of two lobes; each lobe is of a semi-oval form, with their margins furnished with stiff hairs like eye-brows, which embrace or lock into each other, when they close: this they do when they are inwardly irritated.

The upper surface of these lobes are covered with small red glands, each of which appears, when highly magnified, like a compressed arbutus berry.

Among the glands about the middle of each lobe, are three very small erect spines. When the lobes inclose any sub-stance, they never open again while it continues there. If it can be shoved out, so as not to strain the lobes, they expand again; but if force is used to open them, so strong has nature formed the spring of their fibres, that one of the lobes generally snaps off, rather than yield.

The stalk is about six inches high, round, smooth, and without leaves, ending in a spike of flowers.

The flowers are milk-white, and stand on foot stalks, at the bottom of each of which is a little pointed bractea, or flower-leaf.

As to the culture of it: the soil it grows in (as appears from what comes about the roots of the plants, when they are brought over) is a black light mould, intermixed with white sand, such as is usually found on our moorish heaths.

Being a swamp plant, a north-east aspect will be the properest situation at first to plant it in, to keep it from the direct rays of the meridian sun; and in winter, till we are acquainted with what cold weather it can endure, it will be necessary to shelter it with a

bell-glafs,

bell-glaſs, ſuch as is uſed for melons; which ſhould be covered with ſtraw or a matt in hard froſts : by this method ſeveral plants were preſerved laſt winter in a very vigorous ſtate.   Its ſenſitive quality will be found in proportion to the heat of the weather, as well as the vigour of the plant.

Our ſummers are not warm enough to ripen the ſeed: or poſſibly we are not yet ſufficiently acquainted with the culture of this plant.

In order to try further experiments, to ſhew the ſenſitive powers of this plant, ſome of them may be planted in pots of light mooriſh earth, and placed in pans of water, in an airy ſtove in ſummer; where the heat of ſuch a ſituation, being like that of its native country, will make it ſurprizingly active.

But your knowledge of univerſal nature makes it very unneceſſary for me to ſay any thing further, than that I am, with the utmoſt regard and eſteem,

aſſured friend,

and very humble ſervant

JOHN ELLIS.